Metropolitan Water Use Conflicts in Asia and the Pacific

Studies in Water Policy and Management

Series Editor
Charles W. Howe

Published in cooperation with the

 EAST-WEST CENTER

Program on Environment
Honolulu, Hawaii USA

In collaboration with the

 UNITED NATIONS
CENTRE FOR REGIONAL DEVELOPMENT
Nagoya, Japan

*With partial support from the U.S. National Committee
on Scientific Hydrology under U.S. Geological Survey,
Department of the Interior, Assistance Award Nos.
1434-92-G-2243 and 1434-93-G-2373.*

Metropolitan Water Use Conflicts in Asia and the Pacific

EDITED BY
James E. Nickum
and K. William Easter

Routledge
Taylor & Francis Group
LONDON AND NEW YORK

First published 1994 by Westview Press

Published 2018 by Routledge
52 Vanderbilt Avenue, New York, NY 10017
2 Park Square, Milton Park, Abingdon, Oxon OX14 4RN

Routledge is an imprint of the Taylor & Francis Group, an informa business

Library of Congress Cataloging-in-Publication Data
Metropolitan water use conflicts in Asia and the Pacific / edited by
    James E. Nickum and K. William Easter
        p.   cm.—(Studies in water policy and management)
    Published in cooperation with the East-West Center, Program on
Environment in collaboration with the United Nations Centre for
Regional Development
    Includes bibliographical references and index.
    ISBN 0-8133-8779-5
    1. Municipal water supply—Government policy— Asia. 2. Municipal
water supply—Government policy—Pacific Area. 3. Water use—Asia.
4. Water use—Pacific Area. I. Nickum, James E. II. Easter, K.
William. III. Program on Environment (East-West Center) IV. United
Nations Centre for Regional Development. V. Series.
HD4465.A78.M48   1994
363.6'1'095—dc20   94-35213
                                                                         CIP

ISBN 13: 978-0-367-01024-9 (hbk)
ISBN 13: 978-0-367-16011-1 (pbk)

To our forerunner, mentor, and colleague
Maynard M. Hufschmidt

Contents

Figures and Tables

Tables

Foreword

Throughout the world, growing urban demands for water are confronting the increasing economic, social, and environmental costs of developing new supplies. The situation is aggravated by the growing pollution of the cities' traditional water sources by industry, agriculture, and households, by the aging of existing infrastructure, and the lack of coordinated policies among agencies, or between water quantity and quality management.

These problems have become particularly acute in the large, rapidly growing cities of the Pacific Rim. Half of the world's largest metropolises are in the Asia-Pacific region, and even relatively large rivers that flow into urban areas relatively clean leave them seriously degraded by point and nonpoint source effluents. Most Asians still live in the countryside, many of them relying on massive quantities of water for irrigated agriculture.

The present volume argues that in Asia and the Pacific, urbanization presents a fundamental challenge to the traditional ways of developing and managing water. By presenting cases from seven quite different urban areas in the region, it shows that in one way or another the "water economy" is "maturing," driven by the increasing economic and environmental costs of developing new supplies, intensifying conflict between new and old users, and the growing importance of water quality. It concludes with a discussion of alternative approaches to urban water management, applicable to urban areas throughout the world. Indeed, the problems of Beijing and Bangkok do not seem to be qualitatively different from those of Denver or Washington, and the range of possible solutions is similar.

This volume should be of use to a wide range of water professionals and students of water administration not only in the Pacific Region but also around the world.

Charles W. Howe
Series Editor

Charles W. Howe is professor of economics and Director, Research Program on Environment and Behavior, University of Colorado.

Preface

This volume lies at the confluence of many streams. The catchment is at the East-West Center in Honolulu, which for over a decade has explored the problems of managing water and what it touches in an integrated manner. An early product of this exploration was a 1986 volume in this series, edited by K. William Easter, John Dixon, and Maynard Hufschmidt, *Watershed Resources Management*.

The headwaters of this book, to the extent that it is possible to single out one source (or extend the metaphor), may be identified as a 1985–1987 joint project on water resources management in the greater urban areas of Beijing and Tianjin. That activity, carried out by the East-West Center and the State Science and Technology Commission of the People's Republic of China, clearly identified both the growing conflict between urban and agricultural water users and the increasing need to adopt demand management policies to resolve this conflict rather than rely on the traditional supply-oriented approach of building new projects.

The question that followed was whether similar conflicts and management prescriptions might apply to areas in Asia and the Pacific other than north China. Borrowing the concept of the "maturing water economy," originally used by Alan Randall to describe Australia's water situation, we hypothesized that, indeed, certain forces are at work that necessitate a rethinking of how water is managed in this region, and possibly in other areas as well. In particular, Asia's cities are large, growing, and mostly located within the world's most extensively irrigated farmland. The bulk of the water that can be claimed for human use at relatively low expense has already been appropriated; this applies to its quality (assimilative capacity) as well as its quantity. Supply-side options to develop or treat additional water are increasingly expensive both economically and politically, making demand-oriented solutions more and more attractive as alternatives.

The papers in this volume originated in a 1989–1993 project on water use conflicts in Asia-Pacific metropolises, which was conducted by the East-West Center in cooperation with the United Nations Centre for Regional Development (UNCRD) in Nagoya, Japan. For the UNCRD, this project was an outgrowth of its 1987–1990 project on River/Lake Basin

Approaches to Environmentally Sound Management of Water Resources and its deep experience in regional and urban studies. Dr. Kenji Oya, Programme Specialist, was particularly instrumental in promoting the joint activity on metropolitan water conflict, with the support of then Director Hidehiko Sazanami. Collaborating in the project, and providing a beautiful venue for two of its three workshops, in 1989 and 1993, was the Lake Biwa Research Institute, where Dr. Masahisa Nakamura played a key role in establishing and maintaining that connection, and gave valuable advice throughout the project.

From 1991 to 1993, coinciding with the project on water use conflict and with the drafting of the chapters in this volume, the World Bank invited K. William Easter to Washington, D.C., as a principal drafter of its water resources management policy paper. The perspective in that paper is much in keeping with the theme in this volume—that water is one resource, currently managed by separate administrative and professional channels, even within international financial institutions, too often without proper recognition of its economic value, especially across sectors. Our concluding chapter has been considerably updated to incorporate the insights gained by Easter at the World Bank.

A number of people have helped us make the difficult transition from raw manuscripts to a volume readable and visually attractive for publication. In addition to those already mentioned, these include Helen Takeuchi, Joyce Kim, and Linda Shimabukuro of the Program on Environment, East-West Center; and Ellen McCarthy and Kellie Masterson at Westview Press. Special thanks are due to Regina Gregory and Daniel Bauer for their excellent and dedicated service in respectively carrying out the primary and tertiary treatment of the flow of words and figures (end of metaphor).

We are grateful for the financial generosity of the U.S. National Committee on Scientific Hydrology, which underwrote the editing and publication costs of this work as a part of the U.S. contribution to the International Hydrological Programme of UNESCO.

James E. Nickum
K. William Easter

About the Contributors

Michio Akiyama, senior researcher (*shunin kenkyuuin*) at the Lake Biwa Research Institute in Otsu, Japan.

Seiji Aoyama, environmental officer of the Environment Division of the Shiga Prefectural Government; formerly associate expert at the United Nations Centre for Regional Development, Nagoya, Japan.

K. William Easter, professor of agricultural and applied economics at the University of Minnesota.

Francisco P. Fellizar, Jr., assistant secretary of the Department of Science and Technology of the Republic of the Philippines; formerly dean of the College of Human Ecology, University of the Philippines at Los Baños.

Yok-shiu F. Lee, fellow in the Program on Environment at the East-West Center.

James E. T. Moncur, professor of economics at the University of Hawaii, Manoa.

Masahisa Nakamura, associate head (*sookatsu kenkyuuin*) of the research division of the Lake Biwa Research Institute in Otsu, Japan.

James E. Nickum, senior fellow in the Program on Environment at the East-West Center.

Kenji Oya, environmental management planner at the United Nations Centre for Regional Development, Nagoya, Japan.

R. Sakthivadivel, senior irrigation specialist at the International Irrigation Management Institute, Colombo, Sri Lanka; formerly director of the Centre for Water Resources at Anna University in Madras, India.

Euisoon Shin, professor of economics at Yonsei University in Seoul, Korea.

Ruangdej Srivardhana, professor of economics at Kasetsart University in Bangkok, Thailand.

K. Venugopal, assistant professor of civil engineering, Centre for Water Resources at Anna University in Madras, India.

1

The Maturing Metropolitan Water Economies

James E. Nickum and K. William Easter

The towns and cities of Asia and the Pacific are big and expanding rapidly. One-half of the world's megacities are in the region, and "millionaire" metropolises are commonplace.[1] Nonetheless, Asia is the least urbanized region in the world in terms of share of population living in cities.[2] In part because of this, its rate of urbanization, especially in South Asia, is the highest in the world (see Figure 1.1). Since most of the population in Asia and the Pacific is still rural and increasingly mobile, the stresses of the region's urbanization are likely to dominate the political, environmental, and economic scene over the coming decades. Prominent among these stresses will be growing conflicts over water.

The growth of cities has intensified urban-rural interactions. Urban areas are drawing more and more heavily on the food, labor and natural resources of the rural sector, while providing the latter with incomes, employment, lifestyle models, and domestic and industrial wastes. Often, these interactions are mutually beneficial. When it comes to water, however, the direct and indirect demands of urban areas on water quantity and quality are leading to pronounced conflicts.

Because one person's or entity's use of water commonly affects the quantity and quality available to others, conflicts abound. To use an economic metaphor, water is an excellent vector for externalities. Problems of allocation among water use sectors—agriculture, industry, urban water supply and sanitation, fisheries, navigation, hydropower, environmental preservation, and recreation—are becoming increasingly acute as water resources are more fully used and pollution increases.

The conflict between city and farm is particularly acute in the Asia-Pacific region, where most major cities are surrounded by irrigated

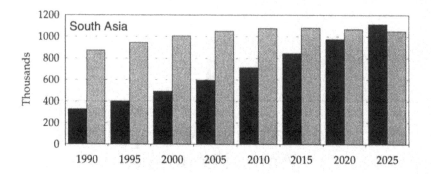

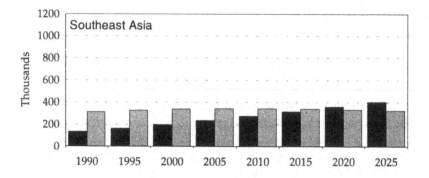

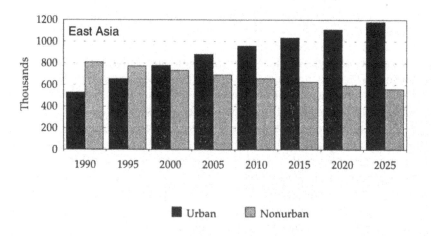

FIGURE 1.1 Urbanization trends in Asia (data drawn from United Nations 1991:123, 125, 135, 137).

agriculture, a "low-value" alternative to "traditional" urban water sources such as groundwater and interbasin transfer, which are becoming increasingly expensive or unavailable. Furthermore, the conflict often becomes critical well before the population of an urban area reaches into the millions, since a significant number of the cities in the region are located in small catchments or on islands.

While intersectoral conflicts such as between irrigation and urban uses of water are increasingly salient, and the main reason we became interested in this problem, they are far from the only problem domain confronting water policymakers. Geographical (upstream-downstream) conflicts between users are still important and are widening in scope as cities reach into other basins for water. Within cities, the problem of serving growing populations settling on extending and increasingly less benign landscapes is laden with strife. In Chapter 2, Y. F. Lee notes a number of sharpening conflicts within the urban water supply and sanitation sector: between tight budgets and high-cost infrastructure, between cost recovery and commitments to provide subsidized "lifeline" services, between system expansion on the one hand and maintenance and repair of existing systems on the other, and between conventional centralized systems and innovative local facilities.

In addition, the conflict between the served and the unserved is salient within almost all use sectors. For example, although drinking water normally constitutes only a small share of total water usage, its quality requirements and direct linkage to health and human well-being have made its provision a central concern of governments around the world. Yet many remain "unserved," left to less sanitary or more costly alternatives to public provision.

In 1980 only about 40 percent of the world's population had access to a safe and adequate supply of drinking water. Coverage was lower in rural areas and suburban areas, and for low-income people wherever they lived. It is now clear that, despite immense efforts during the U.N.-declared Decade for International Drinking Water Supply and Sanitation, by 1990 only a few countries had reached the Decade's goal of providing a safe water supply to all their citizens. Between 1980 and 1990 water supply coverage in the world's urban areas rose from 77 to 82 percent of the population, but was constrained by the rapid increase in the number of city dwellers. In Asia and the Pacific, water supply was extended to 185 million additional residents during the decade, far more than the 148 million remaining unserved in 1980. Yet because of urban growth, urban supply coverage increased by only 4 percent, and the number unserved increased by 27 million, to 175 million. In many cities, even those who are connected to the public supply are not well served in terms of timing, quality, or availability of tap water. The water supply situation in many cities remains difficult, and may be worsening (Christmas and de Rooy 1991).

Water Shortages

Water quantity problems are accelerating in most Asian-Pacific metropolises because of their growing economies and populations. Although, in general, most population growth in these cities is due to urban births, the worldwide trend toward greater mobility of populations is likely to lead to increasing pressure on cities of rural-to-urban migration. Reductions in rural water supply, if they lead to loss of livelihood, would accelerate this migration. Thus cities will be faced with the question of how to deliver water to an ever-increasing urban population without harming those who remain on the farm.

As metropolitan water authorities try to secure added supplies to satisfy their residents, they compete with other sectors for water and for funds. This is reflected in a rapid upsurge in the long-run marginal costs of supply in cities around the world (Munasinghe 1992:13–15). Competition for budgetary allocations will intensify as more funds are requested for improving water treatment facilities, replacing leaky pipes, or developing new surface or groundwater supplies. All these activities cost money,[3] yet in most areas users have not been paying the full cost of their water supply. The larger the subsidy involved, the more government funds will be required to keep up with growing demands.

If the primary source of water for a city is a surface flow, competition will extend up and down the stream. How well each metropolitan area does in this competition will depend on the rights structure and water policy at the national and local levels. Which sector has the highest priority rights to water: domestic, industry, agriculture, recreation, or hydropower? In some countries the ranking is spelled out, almost always giving first priority to domestic users.[4] In other countries priorities are not spelled out.[5] In ordinary times, a first-in-use (e.g., appropriationist) principle is usually applied, although it is often difficult to control subsequent upstream uses, especially but not exclusively their degradation of downstream water quality. Multiple purpose reservoirs often allocate water based on a preset rule. In most countries, ad hoc adjustments are necessary in the face of droughts. These adjustments are commonly made by representatives of the various user agencies. In Japan, water rights (*suiri ken*) are effectively suspended during water shortages.

This causes a great deal of uncertainty for a metropolitan area and may encourage it to claim supplies well in excess of current needs. Since weather has an immediate effect on surface water supplies, water scarcity within many metropolitan areas is likely to occur on occasion unless the supply is unusually large. Even those Asian metropolitan areas in the monsoonal tropics have to deal with dry season shortages. The important trade-off in these cases is the cost of investment in excess or reserve supplies versus the losses due to water shortage.

The water shortages will have different impacts depending on how they are exhibited and how the metropolis responds. Scarcity may show up as low water pressure in the delivery system, as restrictions on supply for short periods during the day, as cutbacks in service to certain areas of the city, or as quantity restrictions on each household or on certain activities. The biggest economic losses occur in extended periods of scarcity if industry (including hydropower) does not get water, although, in agriculture, an entire crop may be lost if water is unavailable during brief critical periods; households may face significant cost increases if they must turn to expensive alternatives to public supply. They may also have large "noneconomic" losses imposed on them when their health risks increase due to reduced use of water for sanitation.

Frequent water shortages, or unreliability or absence of public supply, may impel some users to provide their own water through wells. This can lead to an overdraft that allows saltwater intrusion or irreversible compaction of the aquifer. The result may be loss of a valuable resource for the metropolitan area.[6]

Water Quality

In general, the basic concepts that apply to conflicts over water quantity may be used for water quality as well, although the latter is usually more complex due to the larger number of quality dimensions. Sometimes the two are somewhat directly related. The phrase "the solution to pollution is dilution" was once commonly accepted. If the supply of water goes down, the concentration of a fixed input of pollutant will go up. A common strategy for pollution control is to increase water supplies during low flow periods. This does not work very well if there is a severe drought, however, or if for other reasons there are alternative demands on flushing water. It also cannot be applied to certain toxic pollutants with no threshold effect that tend to accumulate in sediments and biota.

Besides acting as a vector for externalities, water is an excellent medium for disease vectors and toxic materials. Four-fifths of all illnesses in the developing world have been associated with unsafe and inadequate water supplies and sanitation (USAID 1987: 23).[7] Urban, and increasingly rural, industries may damage the countryside's water supply through the discharge of chemicals. "Modern" pollutants such as chemical fertilizers and pesticides from farms may degrade the urban water source, as may more traditional sources such as manure and sediment, which tend to increase with economic development.

"Non-point source" pollution from farms and households is often harder to control than industrial discharges. Sometimes, especially with pesticides and organic wastes, farmers damage their own water supply, because of lack of awareness or training and improper or inadequate labeling of chemicals. Lack of alternatives and low incomes, together with the consequent short time horizons, compound the problem among the poor.

The growth of industrial point-source pollution tends to follow economic growth, at least until an economy becomes mature enough that it can afford to expend funds on cleanup or shift to less polluting knowledge-intensive industries. Governments of many developing economies have encouraged adoption of technologies with the lowest production costs, even when such technologies are highly polluting, in order to compete in the world market. They argue that they cannot afford to cleanup the pollution or require the adoption of a cleaner technology. In effect, they adopt a "dirty now, clean later" approach to economic development. This attitude is becoming increasingly untenable, as the health hazards and consequent economic and social losses posed by water pollution in densely populated areas have become more apparent.

Approach of This Book

The primary purpose of this work is problem setting, to investigate the nature of water-use conflicts in and around a broad sample of metropolitan areas in the Asia-Pacific region, and to lay out some broad alternatives, based on economic theory and on actual practice both within the region and elsewhere, for addressing these conflicts. The eight case studies have been chosen to represent a wide range of conditions in terms of income, location, and water availability (see Table 1.1).

What these cases have in common, we contend, is that they can be analyzed in terms of the "maturing water economy," with consequent implications for water conflict and the need to make qualitative changes in the institutional approaches to water management.

Maturing Water Economy

One way of sorting out these quantity and quality problems, and the management options for dealing with them, is to think of a water management system as a "water economy" that develops from an expansion phase into a mature phase (see Table 1.2).

TABLE 1.1   The Case Studies

City	Income Level of Country	Region	Water availability[a]
Bangkok	Medium	Southeast Asia	Local deficit
Beijing	Low	Northeast Asia	Local deficit
Honolulu	High	Pacific	Mixed[a]
Madras	Low	South Asia	Local deficit
Manila	Medium	Southeast Asia	Local surplus
Osaka	High	Northeast Asia	Local surplus
Seoul	Medium	Northeast Asia	Local surplus
Yahagi (Nagoya)	High	Northeast Asia	Local surplus

[a]Local water deficit is here defined as the difference between precipitation and potential evaporation, neglecting recharge to groundwater (Council of Ministers of the USSR, 1977). A city that has a local deficit may still have adequate water through import from other regions, via either a natural flow (river) (e.g., Bangkok) or artificial diversion (planned for Beijing and Madras). Honolulu, which is both a city and a county, is mixed because the leeward side of the island of Oahu, much of which is built-up, is deficit, while the mountains and windward side are water surplus.

The mature water economy is not a true stage-of-development paradigm. Historically, specific locations have tended to alternate between expansion and mature phases in their water development, due in large part to the lumpiness of new supply projects and technological shifts. The way to stimulate an expansionary "rebirth" from a localized plateau of maturity has usually been to open up a new source (e.g., by tapping a more distant point on the river, by building reservoirs upstream, by

TABLE 1.2   The Expansion and Mature Phases of a Water Economy

	Expansion Phase	Mature Phase
Supply of water	Elastic	Inelastic
Demand for water	Expanding rapidly	Expanding more slowly
Social cost of subsidy for increased water use	Low	Rising
Physical condition of water facilities	New	Many old facilities
Competition between different uses	Low	High
Type of externalities	Drainage	Aquifer depletion; water pollution

Adapted from Randall (1981:196).

constructing another water treatment plant, or by pumping from ever deeper or more remote aquifers). In addition, "post-mature" water economies are appearing, such as in London, where reduced reliance on groundwater is leading to a threat of flooding of underground facilities (Baker 1989).[8] Nonetheless, we contend that, for a variety of reasons, the general case in the cities of Asia and the Pacific is toward a more enduring "global" level of water maturity, a stage that will not easily be passed through by building new supply structures. Reasons for this include not only increases in population and economic development but also the growing number of recognized legitimate uses of water; rapidly escalating costs of new supplies, an increasing number of which have to be transferred from other uses; and nonsustainabilities in existing supplies due to increasing costs of maintenance and repair, depletion, and quality degradation. Physical facilities need to be replaced, competition for water is keen, and the demand for high-quality water is increasing rapidly.

The problems of a maturing water economy show up in a number of different forms. As water quantity and quality problems intensify, complaints increase concerning short supply or contaminated water, international hotels install their own water treatment systems, the well-to-do install their own pumps, and pipe failures are more and more frequent. In addition, when water is supplied intermittently people are encouraged to store water from day to day. However, when a fresh water supply is received, the stored water is dumped down the drain by the people (Unvala 1989:35). This direct loss of clean water would not occur if the water supply were assured on a continuous basis.

The concept of the maturing water economy is useful in the present context in a number of ways. First, it provides a useful corrective to constructs of economic development that focus on output per capita, trade, savings, investment, and the like while ignoring the specific character of the local natural resource base. An economically not-so-advanced area, such as Madras and to some extent Beijing, may have a mature water economy that even constrains further economic growth. Wealthier areas, such as Osaka and Honolulu, may be generally operating without a serious quantity constraint. Still, Osaka has witnessed significant spatial shifts in demand and increasing concern with the quality of tap water while in Honolulu. "The city is well into the rising portion of its long-run marginal cost curve" (Moncur, Chapter 9).

The maturing water economy also provides an alternative to quantitative rules of thumb, such as the Falkenmark criterion of 1,000 m^3/capita/year, which for all its merits does not consider the larger context of system boundaries or the mix of current uses (Falkenmark 1986). More important, by focusing on a qualitative change in the nature of the water problem, it points to a need to adjust the policy options and thereby the institutional rules that govern water management[9] (see Chapter 11; cf. Nickum 1992). In particular, it is necessary to give greater importance to demand management; to move away from an agency-directed, primarily engineering approach; to draw on or defer to the present and potential strengths of stakeholders outside the official agencies, notably water user organizations, private vendors, and non-governmental organizations (NGOs); and to adopt more market and quasi-market approaches,[10] which are more sensitive to shifts in the social and economic value of water. Finally, the maturing water economy affirms the fourth "Guiding Principle" for water management set by the pre-Earth Summit International Conference on Water and the Environment (in Dublin, January 1992): "Water has an economic value in all its competing uses and should be recognized as an economic good" (*Waterlines* 10[4]: 3, April 1992).[11]

Summary of the Chapters

The eight case studies presented in this volume identify a number of different perspectives on the maturing water economy in Asia and the Pacific.

Beijing (Chapter 3)

We begin with an introduction to the growing crisis in China's capital city. A few decades ago, the urban areas of Beijing relied on groundwater, artesian springs, and a small surface diversion from the neighboring mountains. There was little irrigation in the rural areas until after the Miyun Reservoir was completed at the beginning of the 1960s to serve the rural areas of Beijing and downstream Tianjin and Hebei, and tubewells were dug a decade later. As urban demands for water grew with population increases and development of water-using industries (notably chemicals), the city asserted a priority claim over the water of the Miyun. A joint study of medium-term options for Beijing and nearby Tianjin, carried out in 1985–1987 by the East-West Center and the State Science and Technology Commission of China, adapted a methodology originally developed for Denver that places demand options on a par with supply-oriented projects. The chapter by Nickum presented here tracks the evolution of Beijing's maturing water economy and describes the results of this and related studies. These results show the extent of the untapped potential for demand management, possibly including further transfers from agriculture.

Madras (Chapter 4)

Most of the rapidly growing major cities on the Indian subcontinent face serious problems in urban water delivery, both in terms of coverage and, for those fortunate to be connected, in the intermittency of supply. Madras has the lowest per capita drinking water consumption among India's large cities.[12] Like Beijing, it is located in a hydrologically deficit area with only minor surface flows nearby, but unlike Beijing, Madras has no backdrop of runoff-rich mountains. Water shortage has been a major factor in Madras' industrial stagnation and decline in agricultural output. In the city, the delivery system is based on an ancient core operated at low pressure. Public and private trucks supplement the fixed pipe system. Rural impoundments, known as tanks, have been purchased by the city or filled in by its expansion.

Groundwater has been tapped by farmers, suburban industries, and the urban core since the identification of a number of nearby aquifers in the 1960s. The aggregate potential in these aquifers is far from fully exploited but would not be a major source even if fully developed. In addition, localized stresses are already beginning to show. Near the coast, salt water is intruding into the aquifer. In other areas, the farmers rely on the same groundwater that is being pumped out and sold to urban users. The resulting conflicts have led to regulation of private sales of groundwater.

A supply side solution for Madras is to divert the water of the Krishna, via a technically "simple but massive" storage and diversion scheme. If it operates at the planned level, the Krishna project will appreciably alleviate Madras' current water deficit, but only for a few years. Subsequent major diversions may be politically contentious.[13]

At the low levels of water use and high rates of poverty in Madras, some reluctance to adopt pricing methods to regulate demand is understandable. Nonetheless, there does seem to be a widespread willingness to pay that is higher than the current charges, as evidenced by the flourishing private market for groundwater.

Manila (Chapter 5)

Hydrologically (and, perhaps, economically[14]) much better off than Beijing or Madras, Manila shows that reduced reliance on supply-oriented approaches makes sense even in relatively water-abundant but finance-poor areas. The only intersectoral conflict of importance noted in Fellizar's study, written largely from the perspective of the Metropolitan Waterworks and Sewerage System (MWSS), is between urban supply and hydropower generation from the Angat Dam, Manila's principal surface source of supply. Indirectly, but very importantly for any city depending on reservoirs for raw water, another conflict is brewing over the use of the watershed behind the Angat. By reducing the quality of Manila's water supply, uncontrolled watershed development could increase the city's costs appreciably. Similarly, it is increasingly difficult to consider nearby Lake Laguna as a source of municipal supply due to quality degradation, especially in the outlet area near the city.

Despite a heavy debt-servicing burden to domestic and international lenders, a high rate of "nonrevenue" water[15] (61 percent in 1987), and a service coverage of only about one-half of metro Manila's 8 million people, the city's water authority, the MWSS, has succeeded in becoming financially viable. Its approach has been to rehabilitate its undersized and heavily tapped distribution system while extending service to areas that are likely to increase cost recovery. The MWSS has also sought to regulate demand through graduated water rates, but the effectiveness of this approach in reducing overall municipal consumption is limited by the predominance of nonrevenue water in the service area and the extensive reliance on groundwater outside the MWSS system.[16]

Osaka (Chapter 6)

Like Manila, Japan's second city, Osaka, is situated near a large lake (Biwa), which it has drawn on for much of its supply for some time, via the Yodo River. The city's water conflicts have tended to be more of a classical upstream-downstream nature than between different uses. In part this is due to the relative abundance of water[17] in the region, and in part to the small share of land devoted to agriculture in the lower Yodo River basin. The study presented here by Akiyama and Nakamura provides a good introduction to the dynamics of water supply institutions in a rapidly changing city. For example, in 1972, Osaka city entered into a "post-mature" period of declining aggregate demand for water, while suburban water demand in Osaka Prefecture continued to grow, leading to a net stabilization of the region's total demand. The prefecture responded by instituting a coordinated, or "regionalized," water supply system, in large part to overcome problems of access to stable water sources by individual municipalities. According to the authors, regionalization has mitigated conflicts between uses, over groundwater, and among municipalities, but it may aggravate upstream-downstream conflicts.

Seoul (Chapter 7)

There is also little intersectoral competition over quantity in Seoul, as supply augmentation is still possible for the next decade through the construction of four more dams on the Han River. The big problem is quality, due to upstream discharges and a lack of watershed enforcement. This has required the city to import more water than otherwise from the Paldang Reservoir upstream and to face increasing costs of water treatment. Increasing block pricing is enforced, with different rates for different categories of users. Still, this system, which is also used in most of the other cases, is not equivalent to pricing at the margin, as there is considerable cross-subsidization between uses. According to Shin, the price system is also not supplemented by adequate water conservation programs.

Bangkok (Chapter 8)

It may come as a surprise that the city once known as the Venice of the Orient is having water supply difficulties. Even though it is situated in a locally deficit area (annual potential evapotranspiration of over 1,500 mm exceeds precipitation of 1,200 mm), Bangkok is located along the banks of a major river, the Chao Phraya. Nonetheless, in large part due to quality problems, primarily organic matter affecting biological oxygen demand (BOD), Bangkok relies for raw water on a diversion from a barrage at Samlae, about 60 km upstream from the central city, and from groundwater extraction. It shares rights to the water of the Chao Phraya with upstream rice irrigators and fruit growers. The latter claim a flow equal to that of Bangkok and the rice growers combined, in order to prevent seawater intrusion that could destroy the orchards. This puts them in conflict with both rice growers, who do not have enough water during the dry season to grow a full second crop, and with the city, which is seeking new sources to meet growth projections and offset groundwater extraction. Bangkok has had serious problems of groundwater overdraft and land subsidence, intensifying floods. A 1983 cabinet decree directed the Metropolitan Water Authority (MWA) and those industries that could to switch to a surface-based supply system, but implementation is slow, partly because the growth of the city's population and economy has far outstripped the ability of the MWA to keep pace. Bangkok is giving strong consideration to a supply-oriented solution: tapping water from adjacent river basins, using a 100 km long canal.

According to Ruangdej, Bangkok's water-use conflicts are both spatial (upstream-downstream) and sectoral, and over both quantity and quality. In addition, the numerous water laws are often ineffective, due to nonenforcement, low penalties, inadequate updating, and fragmentation. Thailand has no overall water code. Nonetheless, the author notes, there have been some successes, especially in controlling industrial pollution.

Honolulu (Chapter 9)

In his discussion of the Pacific's largest city, Moncur shows the interrelationship between the development of Honolulu's water economy and the institutions that have guided it. Pressure on surface sources, especially by the introduction of sugar cultivation in the nineteenth century, was relieved through a land reform that allowed water rights to be transferred. The development of groundwater, now Honolulu's principal source, beginning in 1879, allowed the island's water economy to enter a prolonged, unregulated expansionary phase for half a century, at which time symptoms of overdraft led the city to establish the Board of Water Supply to operate the city's water system and monitor water development throughout the island, even when done by others. The resulting improvements in efficiency led to a renewed expansionary period, where "raw water remained essentially a free good."

According to Moncur, the *McBryde* decision of 1973, declaring the state to be the owner of all freshwater in the Hawaiian Islands, and ensuing developments, including the 1987 State Water Code, have increased transaction costs (e.g., by prohibiting market transfers between uses and by recognizing new stakeholders) and thereby created a premature condition of maturity. Some aquifers are now in danger of overdraft and have been placed in management zones where well digging and withdrawals are placed under tighter regulation.

Unlike Bangkok, which has adopted a similar approach, Honolulu has few unclaimed alternative sources of water. If free trading were allowed, however, much of the nearly one-half of Honolulu's water used by two sugar companies would probably be transferred to nonagricultural uses. Moncur's indictment of the regulatory approach, especially when it has not grown out of the evolution of the water economy, and his concomitant advocacy of greater reliance on the market may be of broader relevance to many Asian cities, where the state asserts ownership over water and prohibits private exchange.

Yahagi (Chapter 10)

Oya and Aoyama use a different perspective, that of a "suburban" river basin just east of Nagoya. The Yahagi River basin, containing one of the most highly used rivers in Japan, is the site of intense economic activity in most major areas of water use—agriculture, aquatic products, industry (Toyota), hydropower development, and urban expansion. Of these sectors, industry has shown the greatest ability to adjust through recycling, while agricultural use has actually gone up due to a reduction in reuse (because of quality problems). The authors note how established rights of priority favoring agriculture are in practice suspended when a periodic drought threatens other users, especially urban users. One of the strengths of the Yahagi case is that it shows not only how things ought to be but also how they are negotiated in practice.

Management Alternatives

Randall (1981) originally set forth the idea of the maturing water economy in order to argue in favor of a market-oriented approach to water management. Markets have long been accepted as legitimate allocation mechanisms in many of the Western states of the United States aside from Hawaii. Yet most of Asia and the Pacific prohibits or inhibits resorting to market solutions, in part because of the assertion of ownership over water by the state,[18] but also (and underlying that assertion) because of concerns that certain properties of water, notably its public good characteristics, require an alternative approach to growing complexity and conflict. In the final chapter, we explore some of the advantages and disadvantages of alternative institutional approaches to water management, and emphasize the need not to choose between one extreme and the other but to search out an appropriate balance between market, administration, and user involvement.

Notes

1. In 1991, 7 of the 14 urban aggregations in the world with a population in excess of 10 million were in the Asia-Pacific region (rank in parentheses): Tokyo-Yokohama (1), Seoul (4), Osaka-Kobe-Kyoto (6), Bombay (7), Calcutta (8), Manila (12), and Los Angeles (13). Seventeen of the 34 cities with populations over 5 million were in the region. Besides those previously listed, these included Jakarta (15), Delhi (18), Karachi (20), Shanghai (23), Taipei (25), Bangkok (28), Madras (30), Beijing (31), Hong Kong (32), and Pusan (34). Five out of nine additional cities projected to reach 5 million by the year 2000 are in the region: Tianjin, Bangalore, Nagoya, Dhaka, and Lahore (Jamison 1991:A–38 to A–39). With apologies to those left out, we have restricted our case studies of Asia and the Pacific, a very large and heterogeneous area, to northeast, southeast, and south Asia and the Pacific Islands. Our observations and conclusions should apply generally to most of the omitted regions (e.g., Australia and Asiatic Russia), although perhaps not so well to the more arid lands of West Asia and the Middle East.

2. According to Pernia (1991:116), in 1990, 30 percent of the population of Asia and the Pacific was urbanized. In Africa, it was 33 percent, while for Latin America it was 72 percent. Given the wide variation in statistical definition of "urban," it is difficult to say whether Asia is "really" less urban than Africa, and there are wide variations among countries on both continents. Urbanization rates (percent population in urban areas) in developing countries in Asia and the Pacific ranged from less than 10 percent in Nepal to 71 percent in the Republic of Korea.

3. The Preparatory Committee on water resources for the United Nations Conference on Environment and Development (UNCED; the Earth Summit, April 1992) estimated worldwide annual (1993–2000) investment requirements for drinking water and sanitation at $20 billion, sewage collection and treatment at $9 billion, and urban drainage at $9 billion. While these figures are somewhat wishful and apply to more than Asia-Pacific cities, they do show that water-related investments are likely to be a significant claimant on public and private resources. Also, the numbers cited do not include operation and maintenance, where full cost recovery is most imperative.

4. For example, Article 14 of China's Water Law of 1988 gives priority to

urban domestic users, with agriculture, industry, and navigation subordinate (and not ranked among themselves) (*Zhongguo Shuili* [China water resources], 3/ 1988, p. 4).

5. "Under the existing law [in Thailand], there is no mention of priorities because the general assumption is that water is abundant and demand low" (*Bangkok Post*, 26 July 1992; reprinted in *Environmental News Briefing* [Bangkok: ESCAP], July 1992, p. 135).

6. In some cases, even in the absence of transaction costs (for definition, see Chapter 11), wells may be more economical sources of water, in terms of both private and social cost, than a public supply that may require more treatment or longer transportation lines.

7. Public health experts caution about single-factor generalizations. It is very difficult in practice to determine the health effects of poor water quality and, consequently, of improvements in water quality, because of the importance of other factors, including behavioral patterns and the institutional environment (Chen 1983:236).

8. The same source (p. 19) identifies Paris, New York, Tokyo, Liverpool, Nottingham, and "some West German cities" as having similar problems. It should be noted, however, that the abandonment of groundwater is usually due to demand management—increased recycling rates in industry and better integrated water supply systems. In some cases, such as Bangkok, Osaka, and Tokyo, groundwater use has been discouraged due to land subsidence. Thus even "postmature" water economies can be fit into the framework without undue contortion.

9. Following Bromley (1989) and others, we define "institutions" or "institutional rules" as the "rules of the game," such as ownership and use rights, markets and administrative mechanisms, and which may be either formal or informal.

10. Market approaches are discussed in Chapter 11. The term "quasi-market" here means mechanisms such as marginal cost pricing that seek to replicate the efficiency properties of a market.

11. A more global analogue to the maturing water economy is Herman Daly's "full world economics," with its underlying thesis "that the evolution of the human economy has passed from an era in which manmade capital was the limiting factor in economic development to an era in which remaining natural capital has become the limiting factor" (Daly 1991:18).

12. Madras uses approximately 70 ℓ/capita/day, less than one-half that of Delhi. Correction should be made for the availability of nonpotable groundwater which, unlike in Beijing, is still accessible to Madras households from the shallow aquifer. Nonetheless, Madras is widely considered one of the most water-short cities in India, with recent droughts making this painfully evident.

13. A favorable court decision at the end of 1991 allocating additional water from the Cauvery, a river south of Madras shared with Karnataka and Kerala, led to deadly riots in Karnataka (Gleick 1992:11).

14. According to the World Bank (1992:218), per capita 1990 GNP in India was $350, in China $370, and in the Philippines $730. China's GNP is unusually understated in "real" terms, however. GDP/capita in 1985, adjusted to reflect

purchasing power, was $870 for India, $2,472 for China, and $2,168 for the Philippines (Summers and Heston 1991:352–353).

15. "Nonrevenue" water is that portion which enters the delivery system but which does not yield revenue to the water authority. It includes leakage, unauthorized taps, and authorized deliveries that are not paid for.

16. Munasinghe (1992, Chapter 13), details the problem of groundwater depletion in Metropolitan Manila. He notes that the MWSS has identified the largest extractors and is planning to extend distribution facilities to them to replace their supply. This approach has also been adopted successfully in Osaka, but less so in Bangkok (see respective chapters).

17. Lying in a river delta, Osaka has access to both surface water and groundwater. Overextraction of the latter, primarily by industry, caused land subsidence and led to a changeover to surface supplies through industry-based supply systems.

18. India appears to be at least a partial exception. Madras has purchased irrigation facilities from villages for urban supply, although apparently on its own terms.

References

Baker, Geoff. 1989. "The Danger of Rising Aquifer Levels." *World Water* 12(6): 19–20.

Bromley, Daniel W. 1989. *Economic Interests and Institutions: The Conceptual Foundations of Public Policy.* New York: Blackwell.

Chen, Lincoln C. 1983. "Evaluating the Health Benefits of Improved Water Supply," in Barbara A. Underwood, ed., *Nutrition Intervention Strategies in National Development.* Pp. 227–239. New York: Academic Press.

Christmas, J., and C. de Rooy. 1991. "The Decade and Beyond: At a Glance." *Water International* 16(3): 127–134.

Council of Ministers of the USSR, Chief Administration of Hydrometeorological Service and the USSR National Committee for the International Hydrological Decade (1974 in Russian; 1977 in English). *Atlas of World Water Balance,* Leningrad: Gidrometeoizdat and Paris: UNESCO.

Daly, Herman. 1991. "From Empty-World to Full-World Economics: Recognizing an Historical Turning Point in Economic Development," in Robert Goodland, Herman Daly, and Salah El Serafy, eds., *Environmentally Sustainable Economic Development: Building on Brundtland.* Pp. 18–26. Environment Working Paper No. 46. Washington, D.C.: The World Bank.

Falkenmark, Malin. 1986. "Fresh Water—Time for a Modified Approach." *Ambio* 15(4): 194–200.

Gleick, Peter H. 1992. "Water and Conflict," in Occasional Paper No. 1 of the Project on Environmental Change and Acute Conflict. Pp. 3–27. Cambridge, Mass.: Americana Academy of Arts and Sciences.

Jamison, Ellen. 1991. *World Population Profile: 1991.* Washington, D.C.: U.S. Bureau of the Census.

Munasinghe, Mohan. 1992. *Water Supply and Environmental Management.* Boulder, San Francisco, and Oxford: Westview.

Nickum, James E. 1992. "Institutional Approaches for Improved Water Management in a Maturing Water Economy," in Proceedings of *United Nations International Workshop on Water Resources Planning and Management in China*, 2–6 April 1990. [Beijing]: United Nations Department of Technical Cooperation for Development and State Science and Technology Commission of China.

Pernia, Ernesto. 1991. "Aspects of Urbanization and the Environment in Southeast Asia." *Asian Development Review* 9(2): 113–136.

Randall, Alan. 1981. "Property Entitlements and Pricing Policies for a Maturing Water Economy." *Australian Journal of Agricultural Economics* 25(3): 195–220.

Summers, Robert, and Alan Heston. 1991. "The Penn World Table (Mark 5): An Expanded Set of International Comparisons, 1950–1988." *Quarterly Journal of Economics* 106(2): 327–368.

Unvala, B. E. 1989. "Bombay's Water Supply Situation: Drought and Migration Wreak Havoc on a Limited Resource." *Water and Waste Water International*, Feb., pp. 33–37.

USAID. 1987. "Environmental Health and Safety." *The Environment*, Special Report, Fall, pp. 23–30.

World Bank. 1992. *World Development Report 1992: Development and the Environment*. New York: Oxford University Press.

2

Urban Water Supply and Sanitation in Developing Countries

Yok-shiu F. Lee

The growth of population and economic activities in expanding metropolitan regions in the Asia-Pacific, as elsewhere, has led to conflicts over water use between major economic sectors and within each sector. These conflicts are particularly acute because of accelerating urban demands for water and sanitation facilities in developing countries. Although these demands also lead to greater urban-rural conflict, our focus is on conflicts within the urban sector. The average annual population growth rate for cities in low- and middle-income countries worldwide was 6.9 percent between 1980 and 1988, while the overall population in these countries grew by 2 percent (Leitmann 1990). By the year 2000, nearly 40 percent of the 5 billion people in developing countries will be living in urban areas. More than 40 cities of the developing world are expected to have a population of at least 4 million. In 1985, nine out of the twenty largest cities in the world were in the Asia-Pacific region. By the year 2000, twelve of the world's twenty largest cities will be in the Asia-Pacific region, and ten of these will be in developing countries in the Asia-Pacific region (Oberai 1989). The implications of these demographic trends for the provision of social services and infrastructure such as water supply and sanitation facilities are staggering.

Added to the pressure of more people are the relatively high consumption rates of urban populations. In India, Thailand, Malaysia, and Sri Lanka, urban consumers use twice as much domestic water per capita as their rural counterparts. In China, Bangladesh, and the Philippines, urban dwellers use three times as much water per capita as the rural population (WHO 1987). Moreover, urban water supply and sanitation infrastructure are increasingly more costly than comparable infrastructure constructed in rural areas. In many Asian countries, the average per capita unit cost of construction of urban water supply systems is two to ten times more expensive than that of rural water supply facilities (WHO 1987).

All of these trends are intensifying conflicts within the urban water supply and sanitation sector. Prominent among these conflicts are those between the financial constraints faced by municipal authorities and the adoption of high-cost infrastructure standards; between the increasing demand for sector financial viability through full cost recovery and the longstanding commitment by most urban governments to provide subsidized services; between the need for investment in system expansion and the requirement for system repair and routine maintenance; and, finally, between the bias of many managers and engineers toward conventional and centralized systems and the need to design and implement innovative, localized systems to serve the numerous urban residents in slums and squatter settlements.

Before we examine these conflicts in more detail, it would be helpful to have an overview of the symptoms of the major problems permeating the urban water supply and sanitation sector in the developing countries of the Asia-Pacific region.

Symptoms of Major Urban
Water Supply and Sanitation Problems

As noted, the combined effects of urbanization, industrialization, and population growth have greatly increased the demand for potable water and urban sanitation. Yet most urban authorities in the developing world have been unable to provide adequate water supply and sanitation. This has led to continuing unmet demand, low water reliability with poor quality of service and high leakage rates, unrestricted use of groundwater that results in land subsidence and flooding, and exacerbated surface water and groundwater pollution.

Water Quantity

The amount of water actually consumed on a per capita basis varies greatly across national boundaries as well as between the urban and rural areas within a country (Table 2.1). By 1988, 78 percent of the population in the world's developing countries, excluding China, had access to a "reasonably adequate and safe water supply" (WHO 1988a). Although urban areas are generally better served than rural areas, there are considerable inter- as well as intra-urban disparities in access to piped water and in the amount of water actually used by the urban population in developing countries. In Bangkok, one-third of the city's population has no access to public water and has to obtain water from vendors and un-

TABLE 2.1   Water Consumption in the Asia-Pacific Region, 1985 (in ℓ/capita/day)

	Actual Water Consumption	
	Urban	Rural
India	107	40
Indonesia	150	60
Bangladesh	115	30
Thailand	100–150	50–80
Burma	70–110	30–45
Sri Lanka	200	70–170
Maldives	175	175
China	100–200	40–60
Philippines	155	50
Malaysia	230	120–160
Singapore	331	—
Samoa	350	300
Minimum necessary for a healthy life (World Bank recommendation)		30–50

— = Not available.

Sources: WHO 1987; World Bank 1988, Sec. 5.3, p. 4.

treated, often-contaminated sources. In Jakarta, less than one-fourth of the population have direct connections to the municipal water supply system. About one-third of the city's population has to purchase water from vendors at a much more expensive price than tap water (Lindh and Niemczynowicz 1989).

There are some definitional problems regarding the information in Table 2.1. We do not know exactly how each country computes the figures it reports on average daily water consumption, but it is safe to say that there is a wide variety of methods. Some figures may reflect only the amount of water actually supplied to the urban population by the municipal water supply services, as measured by water meters at the household level. If they include water consumed by households not served by the central water supply systems, reporting agencies must rely on estimates of water consumption from nonpublic sources. We need classification of how these estimates are made if we are to have a better sense of actual water consumption in urban settlements of developing countries.

Water Reliability

Continuous water supply at adequate pressures is uncommon in urban areas of developing countries. Intermittent supplies of piped water are the norm in most Asian cities (UNEP 1982). Three reasons are commonly given for this condition: First, conventional means for maintaining water pressure are expensive and technologically difficult. Many developing countries lack the necessary equipment, material, or skilled personnel. Second, power failures are frequent in developing countries. The irregularity of electricity supply not only reduces water pressure by shutting down pumps, it also damages the water pumps and water treatment plants. Third, the shortage of foreign exchange makes very difficult the maintenance of urban services based on imported materials, such as chemicals for water treatment.

Intermittent supplies cause additional problems. They lead to higher peak demands, requiring more expensive, larger diameter pipes (Lindh 1983). The suction created when supplies are turned off greatly increases the risk of pollution intrusion into the water distribution system. Complicating the developing countries' metropolitan water supply problem is their aged piping systems. Post-construction maintenance of distribution systems is rarely adequate. Many new water systems are patched on to existing ones. This may lead to pipe bursts by increasing water pressure throughout the existing systems.

Land Subsidence and Flooding

Groundwater is a major source of water supply for many cities in developing countries. The Bangkok Metropolitan Region relies on deep groundwater for about 50 percent of its total water consumption (Sharma 1986). In Jakarta, two-thirds of the city's population derive their water supply exclusively from groundwater (Sharma 1986). A major problem associated with groundwater use in developing countries is that almost all wells are owned and operated by individual households or firms. As such, groundwater development is commonly unplanned and uncontrolled. For example, the number of wells, amount of extraction, and the depth of extraction in Jakarta remain largely unknown (Sharma 1986).

Unrestricted exploitation of aquifers can lead to unsustainable levels of extraction and to land subsidence. In 1982, groundwater extraction from deep aquifers in Bangkok equaled 1.4 million m^3/day, more than double the estimated safe yield of 600,000 m^3/day (Sharma 1986). Bangkok's piezometric levels have been declining rapidly. The ground level subsided throughout the city from 1978 to 1988, exceeding 70 cm for the decade in the eastern industrial area. While the rate of subsidence appears to be declining slightly, it is still significant (*Proceedings* 1990:159).

The Bangkok region, which suffers from both excess groundwater abstraction and reduced infiltration, is only about 0.1 to 1.5 m above mean sea level (Sharma 1986). Part of eastern Bangkok is already below mean sea level (Sethaputra et al. 1990:55). Land subsidence, exacerbated by poor drainage and high tides, therefore results in a higher frequency of flooding during heavy and moderate rainfalls. Such flooding increases the risk of a range of waterborne and water-related diseases, particularly in the lowest lying and poorest areas.

In Jakarta, as a direct consequence of groundwater overdraft and sea-water intrusion, many wells have been abandoned. Water quality of the city's aquifers has been adversely affected up to 8 km from the coast in northern Jakarta. In parts of central Jakarta, the land surface has sub-sided up to a maximum of 0.8 m. Parts of the core of Indonesia's capital city may be subsiding at a rate of 1 to 3 cm/year (Sharma 1986), and up to 6 cm/year in the northern part (Hadiwinoto and Clark 1990).

Land subsidence carries with it some serious costs such as structural damages to buildings, roads, and railway lines and damages to under-ground pipelines. Moreover, because land subsidence may induce rever-sal of gradients, municipal drainage and sewerage systems can quickly become obsolete, exacerbating other water-related health risks.

In addition, as urban construction creates impermeable or near-imper-meable surfaces, infiltration into groundwater is reduced markedly. Storm runoff is increased and accelerated, and peak flows are enlarged. According to some studies, surface runoff from impervious areas may be hundreds of times greater than runoff from some natural areas. The com-plete urbanization of a watershed may increase the mean flood discharge by five to ten times (Gladwell 1989). Consequently, flooding in low-lying areas is more frequent, exacerbated by increased sedimentation in rivers and canals resulting from urban construction projects and the informal use of channels as a solid waste dump.

Urban Sanitation

In many developing countries, the effort to extend water supply ser-vices to new neighborhoods has far outweighed the commitment to treat and safely dispose of waste in the past decade. The bias may have been an unintentional result of the United Nations International Drinking Water Supply and Sanitation Decade, which set ambitious goals for the extension of water and sanitation services in the 1980s. As shown in Table 2.2, whereas 78 percent of the urban population in developing countries were provided with water supply facilities in 1988, only 66 percent had access to sanitation services. The higher level of developing urban water supply facilities may, ironically, adversely affect the urban environment because increased water supply may lead to a larger discharge of un-treated wastewater.

Household systems such as septic tanks, pit privies, and buckets are the dominant urban sanitation systems for countries in the Asia-Pacific region (Table 2.3). Except for Singapore (and Japan), at most 20 percent of

TABLE 2.2   Water Supply and Sanitation Coverage in Developing Countries, by
WHO Regions (percentage of total population)

WHO Region	Water Supply			Sanitation		
	1980	1988	1990 Estimate	1980	1988	1990 Estimate
Africa						
Urban	69	77	77	57	79	80
Rural	22	26	27	20	17	16
Americas						
Urban	83	87	87	74	81	82
Rural	31	56	62	11	19	21
Southeast Asia						
Urban	67	66	65	29	34	35
Rural	31	56	62	7	12	13
Eastern Mediterranean						
Urban	84	89	90	53	76	79
Rural	31	28	27	8	10	10
Western Pacific						
Urban	75	74	74	92	94	94
Rural	41	50	52	64	67	67
Global total						
Urban	76	78	78	56	66	66
Rural	31	46	49	14	17	18

Source: WHO 1988a.

the urban households in Asia-Pacific countries are connected to central
sewer systems. There are virtually no public sewerage disposal systems
in Bangkok or in Indonesian cities (Lindh and Niemczynowicz 1989). In
Bangkok and Jakarta, human wastes are collected in septic tanks and
cesspools and then discharged into storm drains. This disposal method
may have some health risks, especially in such low-lying cities where inefficient drainage combines with periodic flooding. In addition, septic
tanks and pit latrines tend to overflow during heavy rains. Jakarta's secondary aquifers are widely affected by organic pollution, probably from
this source (Hadiwinoto and Clark 1990).

In China, with more than 200 million city dwellers, there were only 35
small municipal wastewater treatment plants in 1980. As a result, 80 to 90
percent of the estimated 37 billion m³ of sewerage discharged into water
bodies throughout the country was untreated (Liang 1989; Smil 1984). In
Shanghai, China's leading metropolitan region, only 4 percent of the estimated 5 million tons of wastewater discharged in 1980 was treated (Smil
1984). In 1979, 96 percent of the surface water test points in the Shanghai

TABLE 2.3   Water Supply and Sanitation in the Asia-Pacific Region, 1985 (population covered as a percentage of total population)

Country	Water Supply				Sanitation			
		Urban				Urban		
		House Connec-	Public Stand-			Sewer Connec-		
	Total	tion	post	Rural	Total	tion	Other	Rural
Singapore	100	100	0	—	99	89	10	—
Malaysia	96	93	4	76	100	18	82	60
Papua New Guinea	95	80	15	15	99	20	79	35
Korea, Republic of	90	90	0	48	100	9	91	100
Pakistan	83	—	—	27	51	—	—	6
Sri Lanka	82	26	56	29	65	8	57	39
India	76	—	—	50	31	—	—	2
Samoa	75	75	0	67	88	0	88	83
Vietnam	70	—	—	39	—	—	—	—
Nepal	70	—	—	25	17	16	1	1
Thailand	56	—	—	66	78	—	—	46
Philippines	49	35	14	54	83	4	79	56
Indonesia	43	—	—	36	33	—	—	38
Afghanistan	38	18	20	17	5	2	3	—
Burma	36	19	17	24	33	3	30	21
Bangladesh	24	18	6	49	24	4	20	3

— = Not available.

Source: WHO 1987.

municipality were found to be contaminated with heavy metals (Smil 1984). The heavily polluted water could possibly be a contributor to the rapidly rising cancer morbidity and mortality rates in China's leading industrial region, although scientific evidence to support such a claim is not yet reported.

In addition to sewage, improperly discarded solid waste exacerbates the wastewater disposal problem and can be a major contaminant of water sources. Solid wastes dumped into rivers and canals overburden the water's capacity to dilute and flush the waste materials. Left in overfilled and underdesigned waste disposal sites, decomposed solid wastes can easily pollute groundwater through seepage, particularly in the humid tropics.

Sanitary landfills are rare, even in Kuala Lumpur, Manila, Jakarta, and Bangkok (Low 1989). Much of the municipal garbage in developing countries is eliminated by less sanitary means such as open burning and dumping into rivers or canals or into abandoned mine sites and swamp areas. In Bangkok metropolis, at least 25 percent of the garbage produced is disposed of improperly. An estimated 50 to 100 tons of garbage are dumped each day into the city's rivers and canals (Low 1989). In Manila, an estimated 600 tons of solid wastes are left each day on the streets or dumped directly into storm drains, canals, and rivers (Hechanova 1990).

Although improperly handled solid waste can have serious health consequences, solid-waste management, much like wastewater treatment and disposal, frequently receives low priority in the municipal budget. One major reason for this shortcoming is that waste disposal has historically been relegated to the lowest levels of responsibility (Mehta 1982). A related problem is the nonavailability of suitable landfill sites, due partly to high land costs and partly to unplanned rapid urban growth in developing countries.

Key Issues in Planning
for Urban Water Supply and Sanitation[1]

The ensuing discussion presents four arenas of conflict within the urban water supply and sanitation sector: infrastructure standards, cost recovery, unaccounted-for water, and community participation. Conflict in each of these areas is due in large part to institutional factors. Any successful strategy for tackling these problems must consider the possibilities for institutional change. Further research and information needs are noted in the concluding section.

Infrastructure Standards

The availability and quality of urban water supply and sanitation services depend to a great extent on the standards of physical infrastructure systems such as water piping and sewer networks. In many developing countries, there is a tendency to insist on standards that are higher than necessary, sometimes doubling the cost of service delivery. The result is reduced access to water supply and sanitation services (UNEP 1982; Ridgley 1989; Gakenheimer and Brando 1987). Per capita unit costs of providing services have generally continued to increase despite the development of less expensive technologies (Table 2.4). Only a drastic revision of design standards to sharply reduce construction costs is likely to offer hope of providing even minimal levels of public water service to extensive low-income urban neighborhoods.

With few exceptions, the technologies currently in use in developing countries are the same as those employed in developed countries: piped water, full internal plumbing, and conventional waterborne sewerage

TABLE 2.4 Per Capita Unit Costs (median) of Construction of Water Supply and Sanitation System, Asia-Pacific Region (US$ in current year dollars)

| | Urban Water Supply | | | | Urban Sanitation | | | |
| | Home Connections | | Standposts | | Sewer Connections | | Individual Household Systems | |
	1980	1985	1980	1985	1980	1985	1980	1985
Southeast Asia	55	60	—	35	63	81	15	20
Western Pacific	80	96	20	42	220	444	50	73

— = Not available.

Source: WHO 1987.

(Ridgley 1989). Services tend to be provided to those sectors of the population with developed country incomes. Even here there are problems because these systems require expensive equipment and materials, often imported, and trained personnel to operate them, neither of which is easily available in the developing world.

Conventional wisdom suggests that aid-giving agencies and consultants from developed countries encourage the use of costly imported equipment and materials that are produced in their lands of origin. Gakenheimer and Brando (1987) argue, however, that there are strong influences within the developing countries themselves—an "unintentional conspiracy"—that insist on unnecessarily high standards. These include engineers who are most familiar with modern solutions, government agencies who pursue failure-proof and maintenance-free construction, and politicians who wish to avoid being accused of "demodernizing" services. Taken together, these actions and inactions result in an unfortunate tendency toward high and unrealistic standards.

Appropriate standards, and the methods of selecting them, are often more institutional than technological. Because the conflict between financial constraint and the adoption of high-cost infrastructure is caused by numerous actors with different motivations, an overall restructuring of the major institutional relationships may be necessary to yield an effective solution. There is a pressing need to delineate specific, feasible actions that could be effectively taken in different cities in the developing world to make such a rearrangement.

Cost Recovery

Without cost recovery, there is no secure revenue base from which to pay salaries that attract and retain trained personnel. In this and many other ways, the key to improving the performance of urban water supply and sanitation services over the short- and medium-term lies in the ability of public utilities to recover an increasing percentage of the cost of providing services from their customers. Nonetheless, recent data from WHO indicate that with the exception of China, Singapore, and the Philippines, the average water tariffs in the Asia-Pacific region do not even cover the average operating costs (Table 2.5).

At the beginning of the International Drinking Water Supply and Sanitation Decade in 1980, cost recovery was given little attention. By 1985, "an inadequate cost recovery framework" was cited by WHO as the second most serious constraint facing the Decade (WHO 1987). By the late 1980s, many external support agencies and governments in developing countries had reconsidered their cost recovery policies and had started looking seriously for means of implementing cost recovery programs (Katko 1990). By now the question is no longer whether to charge, but how much. One debate taking shape is over whether water supply tariffs should cover only operation and maintenance costs or whether they should also generate resources for future investment. So far very little attention has been paid to cost recovery from sewage services.

In most developing countries, the conventional wisdom is that the poverty of the vast majority of the citizens may make cost recovery difficult. Yet the supposition is that the poor in developing countries cannot afford or will not pay for water services is belied by the widespread practice of water vending at market prices, indicating a high level of affordability and willingness to pay for water by the poor. Nonetheless, it is unclear whether governments have the political will or administrative capability to charge the poor similarly. In many cases, the increasing demand for sector financial viability conflicts with the longstanding commitment by most urban governments to provide subsidized services to their constituencies.

Although water vending is an old tradition throughout the world, little attention has been paid to it in studies of water supply. Recent studies show that, in the absence of access to a public water supply system, people spend substantial amounts of money on vended water (Katko 1990; Okun 1988). Water from vendors costs substantially more than is paid by customers served by the piped water system in the same area (see table, Box 4, in Chapter 11 of this volume). The poor may pay as much as 30 percent of their income for water, whereas the well-to-do pay less than 2 percent (Okun 1988). Supplying free or almost free water to the better-off consumers with house connections therefore often produces very inequitable results.

The success of cost recovery programs relies on, among others, the validity and accuracy of demand estimates for piped water supply. From

TABLE 2.5   Unit Costs of Water Production (operation only) and Water Tariffs, Asia-Pacific Region, 1985 (US$/m³)

Country	Average Cost of Water Production	Average Water Tariff
India	0.08	0.05
Bangladesh	0.09	0.08
Thailand	0.21	0.21
Burma	0.25	0.20
Nepal	0.09	0.07
Sri Lanka	0.25	0.20
China	0.02	0.03
Philippines	0.05	0.15
Korea, Republic of	0.19	—
Papua New Guinea	0.55	0.55
Singapore	0.24	0.29
Afghanistan	0.30	0.15
Samoa	0.09	0.03

— = Not available.

Source: WHO 1987.

the perspective of governments and international donor agencies, the advantages of piped water are so apparent that anyone able to afford tap water is expected to use it. Based on this belief, the proportion of households with income above a certain level has seemed a reliable indicator of future demand. Conventionally, this threshold level has been set at 4.6 percent of household income (van de Mandele 1989). This ignores the availability of alternative sources, however. Consumers may see the picture differently. In Burkina Faso, for example, roughly 85 percent of the inhabitants in the town of Koupela are connected to a new water system at a cost of 20 to 25 percent of calculated monetary income. But an identical system in Boromo—a wealthier place—is only used by one-fourth of the population (van der Mandele 1989). The difference seems to be that unlike Koupela, people in Boromo have a perennial well nearby.

The success of a revenue recovery program also depends on how the revenue is collected. There is an understandable reluctance of many low-income people in developing countries to pay money to a government department that they suspect of corruption or in which they have little or no confidence. To increase the rate of cost recovery, there is a greater need to understand not only how much the user is willing to pay for service but also, from the user's perspective, how the money is to be collected and managed (McGarry 1987). The need to implement cost recovery programs thus leads to a fundamental, but as yet unanswered, question:

What roles should each party in urban water supply development and planning play? The major actors are the central government, the municipal administration, the water agencies, the consumers (household and institutional), and the private supply sector.

Unaccounted-for Water

In many cities of developing countries, about 50 percent of the water that is treated and distributed at public expense is not accounted for by sales (Table 2.6). There is no record of delivery to consumers, and it does not earn revenue for the water supply authorities. World Bank research suggests that, as a rule, if more than 25 percent of treated water is not accounted for, a program to control the losses may prove to be cost-effective (*Urban Edge* 1986). Implementing a formal control policy to reduce both physical (through leakage detection and repair) and nonphysical losses (through improved management practices) typically costs from US$5 to $10 per capita. Studies show that savings and increased revenue will repay this cost within 1 or 2 years (Richardson 1988). Investment to improve the performance of existing assets is thus highly cost-effective. Reduction in unaccounted-for water can allow investments in new works to be deferred or at least reduced in scope, with significant savings. In addition, by improving the system of meter reading and billing or by detecting and charging for illegal connections, revenue can be greatly increased to pay for the costs of treating and distributing the water, as well as the costs of operation and maintenance of the system. Also, if illegal connections are found and charged, it may improve willingness to pay by all. For example, in urban areas in Thailand, each 10 percent of recovered unaccounted-for water would immediately generate an additional US$8 million per year from the 3.5 million people served (Richardson 1988).

Furthermore, where the distribution systems are corroded and broken, appreciable increases in supply do not reach the consumers but result in higher leakage losses. That is, implementation of augmentation projects without controlling leakages could become counter-productive (Kumar and Abhyankar 1988). Passive control of water loss such as repairing leaks only when they are noticed is inadequate. Active control measures such as zone metering are needed to monitor for suspected leaks and systematic leakage detection (Richardson 1988). Leakage from newly constructed systems, which can be as significant as that of old piping networks, can be minimized through careful review of design, materials and construction standards, and tightened monitoring over construction (*Urban Edge* 1986).

Minimizing leakage alone is not enough, however. Nonphysical losses can account for one-fourth (Bangkok) to one-half (Manila) of unaccounted-for water, and they can be reduced at less cost than leakage

TABLE 2.6   Unaccounted-for Water in Municipal Water Supply Systems in Developing Countries

City and Country	Proportion of Unaccounted-for Water to Total Water Supply (in %)	Year	Source
Manila, Philippines	55–65	1984	Richardson 1988
Jakarta, Indonesia	50	1976	Mehta 1982
Mexico City, Mexico	50	1983	UNCHS 1984
Cairo, Egypt	47	1978	Lindh 1983
Bangkok, Thailand	32	1990	Sethaputra et al. 1990

(Richardson 1988). Major strategies for controlling nonphysical losses include the installing, prompt servicing, and recalibrating of meters; the updating and reviewing of consumer records to establish a sound basis for estimating consumption when meters are unserviceable; and the streamlining of bureaucratic procedures to make it easier for customers to make new connections to reduce "theft" of water (Richardson 1988).

Although these strategies and measures are sensible, well-intentioned efforts to reduce high levels of unaccounted-for water, they have met with little success (with notable exceptions such as in Bangkok), indicating the difficulty of solving this seemingly simple problem in developing countries. This also reflects conflict between the apparent need for investment in system expansion and the requirements for routine repair and maintenance of the existing system. Whereas new investment in system expansion usually receives enthusiastic support from managers, engineers, and politicians, routine repair and maintenance work somehow always receive less attention. Hence high rates of unaccounted-for water usually are linked not only to technical problems, but to broader managerial, organizational, and social issues that must also be considered in designing a program to control water losses.

Community Participation

The first half of the International Water Decade was marked by the creation of new technologies and the adaptation of many traditional ones appropriate to the needs of developing countries. Yet both the urban and rural landscapes of the developing world are littered with inoperative pumps that may have been well conceived at the office of a donor agency or country ministry but have fallen into disrepair because of the lack of commitment and participation of the local populations who were purportedly the beneficiaries of such projects (Okun 1988). The need now is not so much for further technological innovation, but for the rethinking of alternative approaches to management and maintenance.

One major reason for failure in water supply and sanitation projects has been that in the minds of international and national planners in this sector, health improvements were the greatest, if not the only, benefits of water supply and sanitation. But the population receiving these services, whether rural villagers or urban squatters, have additional concerns. For them, the reduction in labor spent in collecting water, the prestige of having water in or near the home, or the privacy, safety, and comfort of one's own water closet could be the primary reasons in demanding improvements in water supply and sanitation. Improved health is often a distant third or fourth level of action for many users (McGarry 1987). Well-meaning low-technology alternatives, such as ventilated improved pit latrines, may not be accepted, for example, if the recipient is labeled socially inferior or if the latrines designed do not provide a certain degree of privacy for women. The successful introduction of appropriate technologies requires a thorough understanding of the perceptions of the user communities.

Benefits of community/user participation in all stages of water supply and sanitation projects could include lower costs, greater likelihood of acceptance of technology, and greater user maintenance of the facilities. Studies have indicated that those projects with strong community input are the most successful in terms of reaching the greatest number of the poor with long-lasting services (McGarry 1987). Reorienting project design and implementation methods to incorporate meaningful users' participation is not an easy task, however. It requires substantial structural and attitudinal changes within the implementing agencies. The need to design and implement innovative, decentralized, low-cost communal systems in squatter settlements may also conflict with the bias of many managers and administrators to adhere to conventional municipal systems in urban centers. More understanding is clearly needed of how to motivate the centralized urban water and sanitation agencies to undertake effective outreach measures to low-income communities.

Conclusion

Despite the hundreds of millions of dollars being invested by multilateral and bilateral agencies in extending water supply services to growing cities in the Asia-Pacific region, 40–70 percent of the total water-flow is "wasted" through leaking pipes, non-metered supply, inadequate fittings and appliances, and inappropriate cost recovery programs (Wiseman 1990). Yet these symptoms are merely indicative of the intensified conflicts within the urban water supply and sanitation sector as it comes under the pressure of increasing demand for its services in the wake of accelerated growth of large metropolitan regions. They also exacerbate those conflicts.

Given the enormous gap in supply and demand, the pervasive problems of poor quality of services, and the pollution of surface water and groundwater, any attempt to reliably meet the increasing demand for potable water and sanitation services will require a strategy that considers the major conflicts underlying the many symptoms observed in the urban water and sanitation sector. A successful strategy will need to incorporate the planning and adoption of low-cost infrastructure standards as well as the effective implementation of well-designed cost recovery programs, both of which could greatly enhance the success of project financing. It will also need to strengthen the operation and maintenance of the built water supply systems, particularly the reduction of unaccounted-for water which is a symptom of poor operation and maintenance. Finally, in every major step of the implementation strategy—the planning of low-cost and financially cost-shared systems, and the strengthening of operation and maintenance capacities—the community receiving the services should be involved through active participation to assure the long-term viability of the project.

To extend and improve the quality of services of the urban water sector requires not simply developing more "appropriate" technology but modifying the major institutional and organizational structures to ameliorate the conflicts obstructing the effective mobilization of resources from all major parties. An important but largely untapped source of resources is the users themselves and the communities they comprise. Mobilization of such resources through alternative forms of institutions and organizations would coherently tackle interrelated issues such as appropriate technology, cost recovery, and operation and maintenance.

How could community resources be mobilized to complement those of government agencies to formulate an effective strategy to provide water supply and sanitation services? This query leads to a fundamental question: What role should each of the major parties in the urban water sector play? This question, in turn, leads to others, such as the following: Who should be responsible for community outreach activities to inform users of alternative technologies and the requirements for their construction, maintenance, and operation? How should priorities be set? What are the best methods for determining evaluative criteria and identifying preferences from individual users, communities, and government agencies? What are the best incentives that could be provided to ensure efficient operation and maintenance of decentralized household and communal on-site systems? These are some of the major questions that need to be addressed in searching for an effective strategy to tackle the conflicts underlying the major problems within the urban water sector in developing countries.

Notes

1. The discussion in this section is taken, with slight modifications, from a paper prepared by the author for another book: "Rethinking Urban Water Supply and Sanitation Strategy in Developing Countries in the Humid Tropics: Lessons from the International Water Decade," in M. Bonell, M. M. Hufschmidt, and J. S. Gladwell, editors, *Hydrology and Water Management in the Humid Tropics*, Cambridge, Cambridge University Press, 1993.

References

Gakenheimer, R., and C. H. J. Brando. 1987. "Infrastructure Standards," in L. Rodwin, ed., *Shelter, Settlement, and Development*. Pp. 133–150. Boston: Allen and Unwin.

Gladwell, J. S. 1989. Urbanization, Hydrology, Water Management and the International Hydrological Programme. Paper prepared for Conference on Integrated Water Management and Conservation in Urban Areas, Nagoya, Japan.

Hadiwinoto, S., and G. Clark. 1990. Environmental Profile—Metropolitan Jakarta. Paper prepared for the First Intercountry Workshop of UNDP/World Bank's Metropolitan Environmental Improvement Program, East-West Center, Honolulu, Hawaii.

Hechanova, M. 1990. Profile of the Manila Metropolitan Region—MMR, Philippines. Paper prepared for the First Intercountry Workshop of UNDP/World Bank's Metropolitan Environmental Improvement Program, East-West Center, Honolulu, Hawaii.

Katko, T. S. 1990. "Cost Recovery in Water Supply in Developing Countries." *International Journal of Water Resources Development* 6(2): 86–94.

Kumar, A., and G. V. Abhyankar. 1988. "Assessment of Leakages and Wastages," in *The Proceedings of the 14th WEDC Conference—Water and Urban Services in Asia and the Pacific*. Pp. 23–26. Loughborough, England: WEDC.

Lietmann, J. 1990. Energy-Environment Linkages in the Urban Sector. Working Paper. Washington, D.C.: World Bank.

Liang, Z. 1989. "Water Shortages in Chinese Cities," in *Proceedings of Conference on Integrated Water Management and Conservation in Urban Areas*. Pp. 57–68. International Hydrological Programme, Nagoya, Japan.

Lindh, G. 1983. *Water and the City*. Paris: UNESCO.

Lindh, G., and J. Niemczynowicz. 1989. Urban Water Problems in the Humid Tropics. Paper prepared for International Colloquium on the Development of Hydrologic and Water Management Strategies in the Humid Tropics, Townsville, Australia.

Low, K. S. 1989. Urbanization and Urban Water Problems: A Case of Unsustainable Development in ASEAN Countries. Paper presented at International Colloquium on the Development of Hydrologic and Water Management Strategies in the Humid Tropics, Townsville, Australia.

McGarry, M. G. 1987. "Matching Water Supply Technology to the Needs and Resources of Developing Countries." *Natural Resources Forum* 11(2): 141–151.

Mehta, R. S. 1982. "Problems of Shelter, Water Supply and Sanitation in Large Urban Areas," in *Environment and Development in Asia and the Pacific.* Pp. 227–258. Nairobi: UNEP.

Oberai, A. S. 1989. Problems of Urbanization and Growth of Large Cities in Developing Countries: A Conceptual Framework for Policy Analysis. Geneva: ILO.

Okun, D. A. 1988. "The Value of Water Supply and Sanitation in Development: An Assessment." *American Journal of Public Health* 78(11): 1463–1467.

Proceedings of the National Seminar on Wastewater Management of Bangkok Metropolitan Region with Reference to the Chao Phraya River, Bangkok, 7–8 November (in Thai). 1990. Sponsored by the Secretariat Office of the Coordinating Committee for Royal Development Projects and the International Training Centre for Water Resources Management, Bangkok, Thailand.

Richardson, J. 1988. "Non-revenue Water—A Lost Cause?" in *The Proceedings of the 14th WEDC Conference.* Pp. 147–148. Loughborough, England: WEDC.

Ridgley, M. A. 1989. "Evaluation of Water Supply and Sanitation Options in Third World Cities: An Example from Cali, Colombia." *GeoJournal* 18(2): 199–211.

Sethaputra, Sacha, Theodore Panayotou, and Vute Wangwacharakul. 1990. *Water Shortages: Managing Demand to Expand Supply.* Research Report No. 3 of the 1990 TDRI year-end conference, "Industrializing Thailand and Its Impact on the Environment." Bangkok: Thailand Development Research Institute.

Sharma, M. L. 1986. Role of Groundwater in Urban Water Supplies of Bangkok, Thailand, and Jakarta, Indonesia. Working Paper, Environment and Policy Institute, East-West Center, Honolulu, Hawaii.

Smil, V. 1984. *The Bad Earth: Environmental Degradation in China.* New York: M. E. Sharpe.

UNCHS (U.N. Centre for Human Settlements). 1984. *Environmental Aspects of Water Management in Metropolitan Areas of Developing Countries: Issues and Guidelines.* Nairobi: UNCHS.

UNEP (United Nations Development Programme). 1982. *Environment and Development in Asia and the Pacific.* Nairobi: UNEP.

Urban Edge. 1986. "Tackling the Problem of 'Lost' Water." 10(6): 1–3 (June/July).

van der Mandele, H. 1989. "Resolving Riddles of Price/Demand." *Water Resources Journal,* no. 161: 32–33.

Wiseman, R. 1990. "Low-Cost Technology Favored by UNDP/World Bank Program." *Water and Wastewater International* 5(2): 11–15.

World Bank. 1988. *Information and Training for Low-Cost Water Supply and Sanitation.* Washington, D.C.: World Bank.

WHO (World Health Organization). 1987. *The International Drinking Water Supply and Sanitation Decade, 1981–1990.* Geneva: WHO.

———. 1988a. *Review of Progress of the International Drinking Water Supply and Sanitation Decade, 1981–1990: Eight Years of Implementation.* Geneva: WHO.

———. 1988b. *Urbanization and Its Implications for Child Health.* Geneva: WHO.

3

Beijing's Maturing Socialist Water Economy[1]

James E. Nickum

When Kublai Khan selected the site of Dadu, now Beijing, for China's capital in 1264, he was undoubtedly aware that few locations on the vast, strategic Hai River plain north of the Yellow River were as well endowed hydrologically. He chose well. Beijing has remained the nation's capital for the intervening seven centuries, with only brief lapses from 1368 to 1402 and from 1927 to 1949.

Yet over the past two decades Beijing has faced the problems of a maturing water economy—so much so that it was the inspiration for this study. During the glory days of Beijing's expanding water economy of the 1950s and 1960s, new supply technologies, notably large dams and power wells, encouraged an unsustainable expansion of demand patterns and use rates. Reality set in with a major drought in 1972, which drew attention to the overextraction of groundwater, the reduction of storage in one of the city's two major reservoirs, and a shift in climate away from a relatively wet period. Beijing has responded in a number of ways, most of them concerning demand management. One reason for this is that significant increments in raw water supply would require extensive interbasin transfers of up to 1,100 km.

The purposes of this chapter are (1) to delineate the evolution of Beijing's water economy and subsequent policy responses, with particular focus on the reallocations of existing supplies and the reduction of use rates during the 1980s; (2) to describe briefly the methods and results of studies carried out in the mid-1980s by the State Science and Technology Commission of China in cooperation with the Ministry of Water Resources and Electric Power (now simply the Ministry of Water Resources) and, separately, with the East-West Center; and (3) to note some of the wider implications of Beijing's experience for water-use conflicts in and around other major urban areas in the Asia-Pacific region.

Background

Beijing municipality is quite large, covering 16,800 km² with 11 million residents (as of 1991). This masks wide differences between urban, periurban, and rural areas, however. Administratively, Beijing consists of a small urban core (87 km² with 2.6 million people in 1991), "near suburban districts" (*jin jiaoqu*) (1,283 km² with 4.1 million people), "distant suburban districts" (*yuan jiaoqu*) in the hilly west and southwest (3,198 km² with 1.0 million people), and eight rural counties (12,240 km² with 3.4 million inhabitants). The municipality is located in the middle portion of the Hai River basin on the northeastern edge of the north China plain, just northwest of Tianjin municipality from the Bo Hai (Figure 3.1).

A thumbnail description of the eastern part of China's territory, where most Chinese live, is that the south has lots of water but little land suitable to agriculture and human settlement, while the north has the land but not much water. Beijing is the most favorably situated of north China cities for its water resources. Like Denver, it is located on a piedmont plain, with an average annual precipitation of 625 mm,[2] more than the 565 mm average of the Hai plain or its Colorado counterpart's 516 mm. Its potential evapotranspiration of approximately 900 mm/year, while exceeding precipitation, is among the lowest in the region. Hence the "natural deficit" from local precipitation is the smallest in north China, and is readily compensated for by the inflowing Yongding and Chaobai rivers. Furthermore, Beijing's groundwater recharge from the nearby mountains is the most abundant on the north China plain.[3] Until recently, artesian wells and springs were common in the western suburbs of the city.

Nonetheless, water control in Beijing and nearby areas has long been a major concern of national and municipal governments, even before the Mongol Yuan Dynasty (1271–1368). Year-to-year precipitation varies widely, ranging on record from 242 mm in 1869 to 1,406 mm in 1959.[4] Rainfall also tends to be concentrated in summer storms.[5] The 552 years of record between 1396 and 1948 describe 387 floodings and 407 "relatively large" droughts (Hong and Wang 1992:200). While these events do not appear, in general, to have had a devastating effect on the production and daily life of Beijing's inhabitants,[6] they confronted the state with significant potential risks. The major concerns of the Mongols in siting Dadu were municipal water supply to the world's largest city at the time, flood avoidance, and transportation of grain through the Grand Canal.[7] By and large, those concerns, and the means devised to deal with them, have perdured. While railroads and highways have superseded the Grand Canal, the flooding and water supply problems have continued to confront subsequent governments. At the same time, even post-1949 works

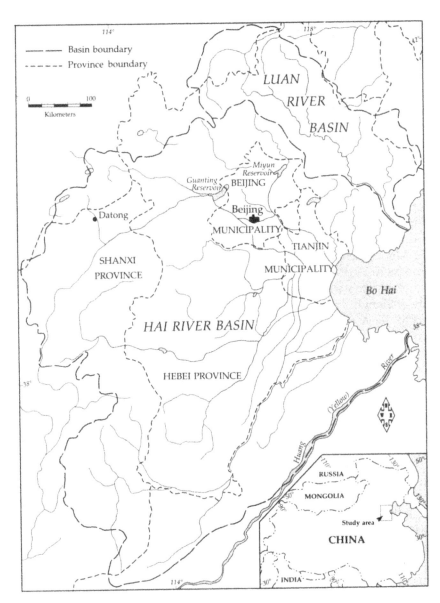

FIGURE 3.1 Beijing and the Hai and Luan river basins.

are often based on plans originally prepared and executed in the Yuan Dynasty. Because agriculture was largely unirrigated, historically there was little water conflict between farm and city, either in quantity or quality (Hong and Wang 1992:198–200).

Beijing's Expanding Water Economy (1950–1972)

The 1950s and early 1960s were a relatively wet period, and claims on supply were modest. Hence, the initial concerns were with avoiding the problems of too much water. Beijing's first large reservoir,[8] the Guanting on the Yongding River, completed in 1954, was intended primarily for flood control, although the water was also used to supply farmers and industries in western Beijing (Liang 1989:51). The municipality's other large reservoir, the Miyun on the Chaobai River, followed during the nationwide push during the Great Leap Forward (1957–1959) to build reservoirs for irrigation. While it was no accident that Beijing's municipal boundaries were last expanded in October 1958 to incorporate the site of the Miyun Reservoir and most of its watershed (Shabad 1972:114), the farmers in northeast Beijing (mainly in Shunyi county) originally shared the reservoir's water with downstream areas of Hebei province and the city of Tianjin.

Problems with many of the poorly considered constructions of the Great Leap, notably salinization of soils on the north China plain, led to a retreat in attempting to harness water for agricultural purposes. A major flood in the Hai River basin in 1963, followed by a wet 1964, only reinforced the sentiment that the problem of excess water at the wrong time was more serious than that of too little (*Huabei diqu* 1985:15–16). The subsequent decade saw a major effort to "harness the Hai River once and for all" (*yiding yao genzhi Hai He*), primarily through the digging of large flood channels through the plain.

Within Beijing, a policy to turn the city from a doctrinally suspect "nonproductive" administrative capital into a "productive" industrial center led to the establishment of a number of water-intensive industries, notably in iron and steel, chemicals, light industry, and textiles. These also required energy from thermal power, itself a major water user. As in most of the world, including the United States, Beijing's economic developers did not consider water a limiting factor in development. Hence they gave little regard to the value of water per se or to the possibility that climate change, either periodic or permanent, or other factors could render unsustainable the use of major new sources.

With the exception of the Shoudu Iron and Steel complex in the western suburbs which can draw on water from the Guanting Reservoir, Beijing's largest industries have relied on their own wells for most of

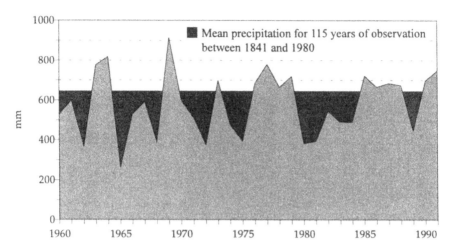

FIGURE 3.2 Annual precipitation in Beijing since 1960 (*Sources:* Data drawn from BJSTN87:345, BJSTN88:171, BJSTN89:145, ZTN90:10, ZTN91:9, and BJTN92:105. Data for 1960–1985 provided author as background data for the Water Management in North China Project (CPR/88/068), United Nations, State Science and Technology Commission of China, and the Institute for Water and Hydraulic Research. Li et al. 1987:165 show a somewhat different pattern).

their water supply (e.g., BJTN92:360). Similarly, the water supply for nonindustrial "domestic" uses has come primarily from beneath the feet of the city's residents.

Beijing's urban water use, industrial and domestic, increased by forty times in 35 years (1950–1985) (*Huabei diqu* 1985:16). Nonetheless, the greatest demand in the municipality was and remains for irrigation water.

Signs of Maturity (1972–1980)

After 1964 came a string of dry years (with the exception of 1969—see Figure 3.2), culminating in the major drought of 1972. South of Beijing, in Shandong and Henan provinces, diversions were reopened from the Yellow River. Some ideas of the Great Leap Forward, like moving large amounts of water across a flat, salinization-prone plain, seemed not so crazy after all. At the same time, throughout north China, a major new water source was opened up. Pump wells, powered by diesel or, in favorably located areas, by electricity, tapped the aquifers underlying the extensive littoral plain. As elsewhere on the plain,[9] the area under irrigation expanded rapidly in Beijing during the early 1970s (Nickum 1993).

While the farmers and food grain planners of Beijing and downstream

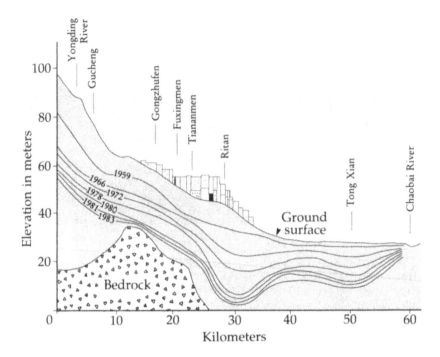

FIGURE 3.3 Groundwater table in Beijing urban district (Gucheng to Tong Xian).

areas were elevating their expectations, it was already clear that urban Beijing was stressing its sources. In particular, a rapid drop in the water table, in some places (e.g., Jiuxianqiao in northeast Beijing) over a meter a year (*Huabei diqu* 1985:39), with a citywide average of 0.5 m/year in the early 1980s, indicated that pumping had exceeded long-term replacement levels, while flows into the Guanting Reservoir from the parched upstream areas were dwindling rapidly. With the decline in discharges for flushing and the rise in urban population, living standards, and industrial production, the quality of the water flowing out of urban Beijing declined as well. Total wastewater (*wushui*) discharged, almost all untreated, increased from 397 million m³ in 1970 to a peak of 728 million m³ in 1980. A significant proportion of this wastewater (35 percent in 1983) was used in downstream irrigation, together with a nearly comparable amount of non-"waste" industrial tailwater. During this period, domestic wastewater increased from 37 to 48 percent of the total discharge (Chen Shenyi, Li, and Long 1986:12, 34–35).

Figure 3.3 shows the dramatic decline in the water table in and near Beijing. By 1980, aggregate overpumping reached 210 million m³ and the aquifer in the western suburbs had been pumped down to bedrock.[10] A

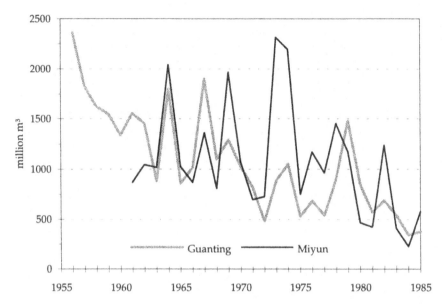

FIGURE 3.4 Inflows into Beijing's major reservoirs (*Source:* Chen Shenyi, Li, and Long 1986:28–29).

pronounced "cone of depression" showed up in the southeastern part of the city where the chemical industry is located. The depletion of the aquifer caused the land surface to subside in 1,000 km² of central Beijing.

The Guanting Reservoir has succeeded marvelously in its original purpose of controlling floods. It has fared less well over time in its acquired duties of supplying water because of the reduction in its inflow (Li Xianfa et al. 1987:161). As shown in Figure 3.4 (based on data in Chen Shenyi, Li, and Long 1986:28–29), the decline of inflow to the Guanting has been erratic but inexorable, with the steepest drops in the late 1950s and early 1970s. The inflow into the Miyun, which has very little upper watershed development, has declined comparably. Hence climatic factors appear to have played at least as large a role as anthropogenic ones in the decline of supply into both reservoirs, at least since 1960.

Nonetheless, increased withdrawals on the upper reaches of the Yongding River in Shanxi and Hebei provinces have had a significant effect on the actual storage in the Guanting. Most of the water withdrawn along the river has been used in agriculture. The upstream irrigated area doubled from 200,000 ha in the 1950s to 400,000 ha in the 1970s (Water Resources and Hydropower PDI 1989:137).

The uppermost reaches of the Yongding system in Shanxi also comprise one of China's major coal-producing regions, including the city of

Datong. Nonetheless, the bulk of the limited water in that district has gone to agriculture, while the coal has gone unwashed.[11] Between Shanxi and the Guanting Reservoir lies the Zhangjiakou Prefecture of Hebei,[12] where about 160,000 ha were irrigated from the Yongding tributaries in 1980 (Rural Socioeconomic Statistics Section 1989:12–13, 30–32).

China does not appear to have any effective mechanism like the inter-state compacts of the United States for establishing and enforcing the water rights of lower vis à vis upper reaches (Johnson and DuMars 1989:352–353). Neither does a market or quasi-market arrangement appear to have been viable. Hence increased withdrawals by upstream areas are factored into downstream plans in the same way as a state of nature. It is estimated that inflows into the Huang-Huai-Hai Plain (consisting of the plains north and south of the Yellow River) will decline by 27 percent, from 56.7 billion m^3/year to 41.5 billion m^3/year, between 1985 and the year 2000. The projected decline for Beijing, of 22 percent, is actually relatively moderate, due presumably to the already reduced level from the Yongding and to its control over the other major watershed, the Chaobai (Water Resources and Hydropower PDI 1989:137).

Adopting Measures (1980–1985)

Chen Jiaqi (1989:1), a top-level engineer in the Ministry of Water Resources, tidily distinguishes the 1970s from the 1980s: "Everyone began to pay attention to the water shortage in north China in the 1970s. Although some measures have been adopted in the 1980s, the problem has still not been resolved very well." Here we discuss some of the measures that were adopted.

As described in the opening chapter of this volume, an expanding water economy relies on new supplies of raw water to meet (or create) demands, while a mature water economy must focus on reallocation and/or reducing the rates of water use in different sectors—so-called "demand management."

Figure 3.5 provides a breakdown of water use by sector and by source (surface and groundwater) for 1980–1984. The general patterns that developed in the 1970s are apparent in these figures, especially for 1980: agricultural uses are dominant, drawing on both surface and groundwater sources; then industry, with a similar pattern; and urban domestic, the smallest use but with stringent requirements of quality and timing, relying virtually entirely on groundwater.

Pressures on the water supply, notably population, living standards and economic growth and drought, intensified following the economic reforms of 1979. "Permanent" urban population, which increased from 4 million to 5.1 million during the 1970s, grew to 5.7 million in 1985, while

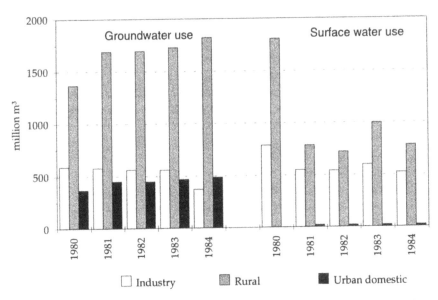

FIGURE 3.5 Beijing water use by sector and by source, 1980–1984 (*Sources: Huabei diqu* 1985:22 [1980]; Ministry of Water Resources and Electric Power 1989:142 [1981–1983]; Li et al. 1987:162 [1984]).

"temporary" population (such as business travelers, resident foreigners, tourists, and ex-farmers without proper residence papers) increased significantly (BJSTN86:154). Industrial growth slowed down only slightly during the Sixth Five-Year Plan (1981–1985), to 7.6 percent per year (measured in gross industrial output value, adjusted for inflation) (Beijing Municipal Statistics Bureau 1987:112–113). Rainfall continued at below normal.

In addition, funds to the Beijing Municipal Water Bureau were cut at the beginning of the 1980s, and 20 percent of the management personnel were released. The remaining government subsidy of over 2 million yuan/year was insufficient for renovations (Hu and Nie 1992:3).

In response to these stresses, Chinese authorities adopted an approach using both supply-and-demand management measures, although sometimes on an ad hoc basis. These included reallocating water use from downstream areas, cutting much of the surface supply to rural areas, reducing agricultural, industrial, and domestic consumptive use rates, and carrying out background studies that would allow a more systematic approach in the future.

1. *Reallocation from downstream.* In 1979, construction began on a separate reservoir on the Luan River to supply Tianjin municipality

downstream from Beijing. In 1981, before this reservoir was completed, the State Council (China's legislature) allocated the water of the Miyun Reservoir water entirely to Beijing, leaving Tianjin to rely on two expensive emergency diversions from the Yellow River.

2. Reduction in rural surface supply. Rural water use fell substantially and permanently in 1981, due to a reduction in surface supply, notably from the two large reservoirs, which had drawn down their reserves in the very dry (P = 95 percent) 1980. The countryside shifted both relatively and absolutely toward groundwater use. The amount abstracted from the aquifers increased from 983 million m^3 in 1979 to 1,691 million m^3 in 1981. Because of the lower transmission losses of wells, the reduction in net water use was less. In addition, the direct linkage between well use and costs to the user undoubtedly encouraged frugality in application.

During this period, "sideline production," rural industries commonly located within the villages, more than doubled as a share of agricultural output value, from 24 percent in 1980 to 54 percent in 1985. In addition, fish farming increased rapidly over the same period (Nickum 1987). Water use figures were not collected for rural industries at this time; their greatest impact may have been on quality, as there was a tendency to relocate the most polluting operations (e.g., paper making and chemicals) from the cities into the countryside (*Huabei diqu* 1985:62).

3. Reduction in urban use rates. A major campaign was initiated to reduce industrial water use rates. First, the city government identified the 259 largest users, those drawing over 100,000 m^3/year, and beginning in 1981 targeted them with "planned water supply measures," presumably involving metering of their own-use wells. The following year it introduced a steeply rising block-rate water-fee system. Use rates dropped markedly. By 1985, only 197 enterprises used over 100,000 m^3, while the minimum for inclusion under planned water supply had been lowered to 36,000 m^3/year, encompassing 350 users and 70 percent of industrial water use. An important element in this water-saving campaign was a "water balance audit" (*ceshi*). The water recycling rate increased from 61.4 percent in 1980 to 72.3 percent in 1985 (Han and Li 1986:182–184) and total industrial water use declined while output increased by 44 percent in real terms (Figure 3.6).

At the same time, water meters were installed in about 600,000 residences (Chinese Research Team 1987:92). While these devices may have served to restrain the growth in home water use during a time of rising income and shifts toward more water-using lifestyles, their impact on Beijing's urban water demand was limited by the relative insignificance of residential use—only 27 percent of domestic demand in 1984 (1.07 out

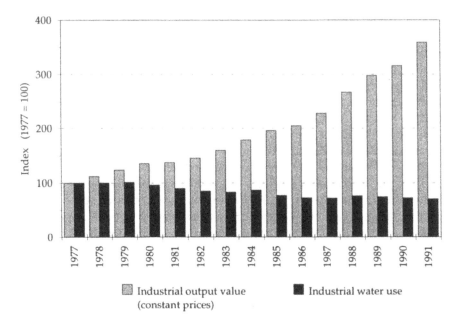

FIGURE 3.6 Beijing's industrial output and water use, 1977–1991. Water use figures for 1985, 1986, 1989, and 1990 are interpolated (*Sources:* Ministry of Water Resources and Electric Power 1989:141; BJSTN88:374; BJSTN89:324; BJTN92:125, 374; ZSN91:295).

of 4.02 million m³). The bulk of domestic demand was attributed to public facilities, both noncommercial (1.98 million m³, or 42 percent) and commercial (0.72 million m³, or 18 percent), with the remainder going to construction, parks, sanitation, and street washing (Chinese Research Team 1987:3–12).

4. *Research*. With the growing water shortage in north China firmly on the agenda, attention also turned to providing a solid research and analytical base for framing policy choices. State-of-the-art modeling and analytical methods were explored, with varying degrees of usefulness.

An important initial step was a comprehensive assessment, grounded in hydrology, of the status of water use. The State Science and Technology Commission of China (SSTCC) funds a number of key science and technology projects in each Five-Year-Plan period. During the Sixth Plan (1981–1985), one of these was the "Assessment of Water Resources in North China and Research on Their Development and Use," carried out largely by the Ministry of Water Resources and Electric Power (now Ministry of Water Resources [MWR]).

This study surveyed water supply and demand conditions in a number of critical subregions in north China, one of which was the Beijing-Tianjin-Tangshan area. Supply gaps were estimated for normal (P = 50 percent), dry (P = 75 percent) and drought (P = 90 percent) years under current conditions and projected to the year 2000.

The potential for water saving was explored, in the case of irrigation with reference to the results of field experiments. Beijing's industrial recycling rate was assumed to increase from 70 percent in 1984 to 80 percent in 2000, but total industrial water use was projected to increase significantly due to growth in output.

Recommendations for overcoming projected shortages included adopting water-saving techniques in agriculture, considering water conditions in industrial siting, and recycling more wastewater. Although all of these are demand management measures symptomatic of a maturing water economy, the study placed greatest emphasis on a supply-oriented solution. The *aqua ex machina* would come in the form of one or two multibillion dollar, 1,100-km diversions from the Yangtze River, termed the *nanshui beidiao* (south-to-north water transfer) projects (*Huabei diqu* 1985).[13]

The MWR has consistently stressed the ultimate need for these diversions (e.g., Chen Jiaqi 1989). Perhaps because of this strong (but evolving) agency orientation toward engineering solutions, or perhaps because of the preliminary nature of its report, the options presented by the MWR were far from comprehensive or detailed, nor were they really compared against each other to allow a systematic comparison of specific alternatives. This was done in the subsequent East-West Center–State Science and Technology Commission of China (EWC–SSTCC) study described in the next section.

Another key national research project of the Sixth Five-Year Plan, begun in 1983 under the sponsorship of the Ministry of Urban and Rural Construction and Environmental Protection (MURCEP), was on the "Urban Ecological Systems of Beijing." This project, which ran for 3 years, placed even greater weight on demand management. It included a water input-output model for Beijing, a system dynamics model, and an analytical hierarchy study (discussed in the next section). The input-output table included eighty-three production sectors, and wastewater discharge included twenty-one quality parameters.

Some of the results were not surprising: agriculture and heavy industry dominated water use. Nonetheless, a more complete picture was drawn of total water use, indirect as well as direct, for specific production sectors. In particular, food, machinery, metallurgy, and medicinal products were found to be heavy indirect users of water quantity, while paper production was identified as the single most significant industrial polluter (Xie, Nie, and Jin 1991). The main values of this study appear to have been the basis it provided for demand management, notably by allowing the targeting of efforts to reduce industrial demand, and its inclusion of quality parameters and price variables.

A system dynamics model was also developed in this study (Nie, Su, and Qin 1986). In theory, models such as these have some appeal in their comprehensiveness and are sometimes used in academic training exercises outside China. They link subsystems for various sectors and for quality considerations into an integrated whole. Nonetheless, their computational and conceptual complexity (the Beijing model had over 200 parameters and variables) and data requirements limit their usefulness in generating clear ("robust") solutions for practical policy choices.

More Recent Developments (1986–1992)

1. *Enter the East-West Center.* A 1985–1987 joint EWC–SSTCC study on water resources management in Beijing and Tianjin examined a wide range of supply-and-demand management options, especially to close the short-term gap (to the year 2000).[14] This was based on the very practical methodology developed in the 1970s for Denver, Colorado (USA), by Milliken and Taylor (1981). The study found considerable scope for use of various demand management options to reduce overall water demand. Supply augmentation was also an option, but to a lesser extent in the absence of long-distance transfers.

Table 3.1 presents a matrix of demand-management options from the joint study. The various technical options are grouped by the three major use sectors: agriculture, industry, and domestic. They range from changes in crop mix and increased irrigation efficiency to increased industrial recycling and domestic leakage reduction programs. For each technical option, the matrix indicates the appropriate policy options that can be used to implement desired changes. These policy options range from pricing policies to various forms of mandatory controls—regulations, quotas, or codes. The table also indicates potential quantities of water saved and the quality of the water. The level of cost of each option and degree of difficulty to implement it are also indicated.

When both demand-and-supply management options were considered, the EWC–SSTCC study estimated total reduction in the projected water deficit in the year 2000. Figure 3.7 shows these results. As seen, in an average rainfall year the "gap" is completely closed and Beijing has a slight surplus. In 1 year in 4 with low rainfall, there is a small, final deficit. In 1 year in 20 of major drought (the 95 percent probability graph), there is still a sizable deficit but one that is much smaller than in the case without use of demand-and-supply management options. More complete

TABLE 3.1 Demand-Management Options Matrix

	Policy Options					Characteristics of Water Saved			Characteristics of Options		
	Pricing[a]		Mandatory Controls[a]								
	Volume	Flat Rate	Regulations	Quotas	Building Codes	Potential Quantity Saved[b]	Quality[c]	Fungibility[c]	Level of Cost[c]	Difficulty of Implementation[c]	Who Pays[d]
Technical Options											
AGRICULTURE											
Reduce effective irrigated area	**	*	**	*	—	200–400	L(SW)	L-M	L-M	H	I,C
Change of crop mix	**	*	**	*	—	150	H(GW)				
Increase irrigation efficiency	**	—	—	*	—	100	L(SW)	M	H	H	I,C
Reallocate water from irrigation to rural industry	**	—	*	—	—	—	H(GW)				
INDUSTRY											
Change industrial product mix	**	—	**	*	—	100–200	Varies	H	M-H	M-H	C
Adopt water-efficient industrial technology	*	—	**	**	—	50	Varies	H	H	M-H	C,G
Increase recycling	*	—	**	**	—	200–600	—	H	L-M	M	C,G
DOMESTIC											
Water-saving devices	*	—	*	—	**	30–40	H	H	H	H	I,C,G
Leakage reduction	*	—	**	—	*	10–15	H	H	M-H	M	C,G
Domestic recycling	—	—	*	—	**	50	H	H	VH	H	I,G

a * = effective   c L = Low   d I = Individual user
** = very effective   M = Medium   C = Collective/factory
— = not effective   H = High   G = Government
b 10^6 m^3 (Beijing)   VH = very high
SW = surface water
GW = groundwater

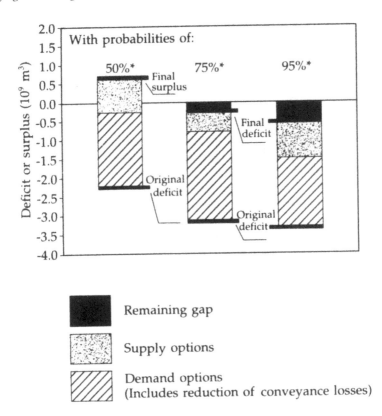

*Probability that the indicated deficit will not be exceeded or that the indicated surplus will at least be achieved.

FIGURE 3.7 Results of supply-and-demand management options, Beijing 2000.

details are given in the East-West Center report on the project (Hufschmidt et al. 1987), the project's Joint Summary Report (EWC–SSTCC 1988), and Fallon and Dixon (1988).

Although the EWC–SSTCC study was restricted to a time horizon of the year 2000, it appears to have had an impact in putting demand management on an equal footing with supply expansion as a policy option. Even those who, like Chen Jiaqi (1989), continue to advocate the *nanshui beidiao* interbasin transfer do so on the reasonable grounds that, like supply-oriented solutions, demand management is subject to rapidly rising costs.

2. Related studies. One difficulty of the EWC–SSTCC results shown in Table 3.1 is that it does not provide a weighting of the different dimensions—not just quantity and cost, but also quality, ability to implement, and cost allocation. One way of providing these weights is through an analytical hierarchy approach (AHP), which derives them from the responses of key decision-makers

An AHP application, part of the 1983–1985 MURCEP study, involving sixty-four experts from over thirty agencies in Beijing who considered twenty-four options, found the greatest support given to two local supply-side approaches (the construction of a reservoir in southwestern Beijing and further exploitation of groundwater), followed by wastewater reuse, the *nanshui beidiao* interbasin transfer, and demand reduction in industry and agriculture. Decidedly less popular, in order, were centralization of water management and planning, regulation of cropping and industrial structures, cutting back on irrigation, restricting industrial growth, and controlling water use of rural enterprises (Xie, Rosso et al. 1989).[15]

Although this method cannot overcome all surveying or informational biases, it does provide a strong sense of what is deemed administratively feasible, especially by administrators. The AHP approach takes the organizational and institutional structure as given. Neither it nor the other studies during the 1980s directly addressed the critical but contentious issue of institutional change.

3. Supply-and-demand management since 1985. Despite continuing rapid economic growth, the crisis situation has abated somewhat since 1984. Beijing's precipitation has increased significantly and has been relatively consistent from year to year. The cumulative effect, shown in Figure 3.8 by a 5-year moving average, has created a marked turnaround from the previous years of decline.

As shown in Figure 3.6, industrial water use has continued to decline while real output has nearly doubled in this period. The primary reason for this has been reduced waste and an increase in recycling rates. Industries have no doubt been motivated not only by policy initiatives such as pricing and quotas but by uncertainty over supply. Major users, such as chemicals, iron and steel, construction materials and textiles, rely heavily on their own wells and nearby surface flows. The decline in the water table and growing intermittence and quality problems of rivers undoubtedly signaled a need to be more frugal and self-contained in water use. In addition, municipal supply to industry declined from 140 million m^3 in 1984 to 118 million m^3 in 1991, apparently to provide more to "domestic" uses.

Unlike industry, domestic users have come to rely primarily on municipal piped water. Their supply of tap water increased from 239 mil-

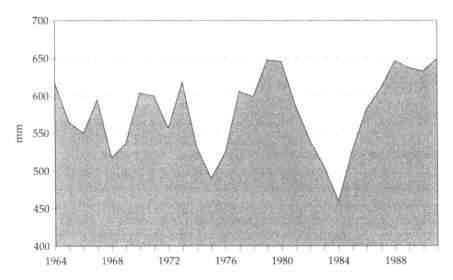

FIGURE 3.8 Beijing precipitation (5-year moving average).

lion m³ in 1984 to 346 million m³ in 1991. As noted previously, about three-fourths of "domestic" uses are nonresidential. Figures on nonresidential domestic uses are not yet available for years after 1987. Comparisons of that year's numbers with those of 1984 indicate some shifts among different kinds of users, however. In particular, education and research facilities reduced their water use, while military and commercial uses increased (ZSN91:303).

Water-saving irrigation, notably using sprinklers, was promoted in this period, especially in areas such as Shunyi county which were heavy users of Miyun Reservoir water. The state and the village collectives provided the bulk of the 400 yuan/*mu* (6,000 yuan, or roughly US$1,000/ha) investment, with the farm households contributing only 27 yuan/*mu* in 1987 and 35 yuan/*mu* in 1988 (field notes, 1989).

Still, there is some indication that total rural use did not decline during this period. This condition may have changed with the completion in mid-1990 of a diversion project from the Miyun Reservoir to the western suburbs, allowing the realization of the 1984 State Council authorization to transfer that source's water from irrigation to urban purposes (ZSN91:402).

The Miyun diversion, dubbed the *dongshui xidiao* (east-to-west water transfer), appears to have been the only significant supply-oriented project built by Beijing during this period. In part this may be due to the effectiveness of demand management, intentional or not, and to the relative abundance of supply. Another factor is a continuing budget constraint inhibiting new projects, especially large interbasin transfers such as *nanshui beidiao*.

In the early 1990s, Beijing's built-up area (urban core and "near suburban districts") discharged about 800 million m³ of wastewater, a modest increase of about 10 percent over the 1980 figure of 728 million m³. Only about 7 percent of this discharge was treated. Even with the quadrupling of treatment capacity with the completion of the Gaobeidian plant in the early 1990s, the bulk of Beijing's wastewater will be discharged untreated for the foreseeable future (BJTN92:88, 506).

4. Prospects. Even with some success in lowering use rates, Beijing's maturing water economy will face a number of increasingly complex and interrelated problems in coming years. The groundwater is being overdrawn, especially in the rural areas. Continued declines in inflow from upstream, especially to the Guanting Reservoir, are probable. Water quality remains poor, especially downstream from the built-up area. All of these problems will be aggravated severely during the next string of dry years.

Thus, China's water planners are continuing to look to the possible need for not one but two diversions from the Yangtze, along the middle route from Henan in addition to the eastern route from Jiangsu. It is unlikely that the state will provide financing for either from its chronically tight budget in the near future, however, unless compelled to do so by a clear and dramatic crisis.

Chinese analysts point to a number of abiding economic and institutional problems as well. Water fees are considered too low, only 0.35 percent of production costs for Beijing's industry (Ren 1991:67). In addition, administrative fragmentation and fiscal rules contribute to a separation of the incidence of costs from projected benefits, inhibiting new investment, especially in conjunctive uses of surface and groundwater (Xie Mei, n.d.).

Implications

It is not surprising that a socialist system, even one that is rapidly reforming, would face a maturing water economy. Indeed, given the penchant for Soviet-style socialism to generate "shortage economies" in general, it would be unusual if Beijing's demand for water did *not* overreach its supply. What may be surprising is the early adoption of demand-management techniques, even before the decision-making methodologies and analytical frameworks were developed to legitimate them.

The most effective of these appear to have been those that relied either on command-and-control approaches or on self-initiated responses, especially by industry, to reductions in the quantity or quality of available water. The former has worked best when the number of control points is limited, notably in the apparently effective identification, auditing, monitoring, and cajoling of the largest industrial users, than progressively widening the span of control.

The strength of Beijing's water economy in making small-number, big-volume adjustments also shows up in the sometimes drastic reallocations of water rights from the Miyun Reservoir which, at the very least, would have led to decades of litigation in the United States. These rights could be transferred readily, in part because of Beijing's unique position as the national capital and home to the leaders who approved the transfers, and because in China there are no equivalents of the binding interstate compacts of the United States. The downside of this for Beijing is that it is virtually impossible for it to exert control over water use upstream of the Guanting Reservoir.

Although metering and pricing approaches have also been adopted in Beijing, especially for water supply to households and large industries, they have had limited effectiveness. As in most of the other cases in the Asia-Pacific region examined in this volume, water markets, either for supply or for pollution rights, are not allowed in Beijing.

Despite much discussion of long-term planning and the necessity for solving Beijing's water problems through long-distance interbasin water transfers, in practice, most of that city's water policy changes have been incremental, multifaceted, and often in response to the most recent crisis. The imperatives of the maturing water economy have led Beijing to move away from a heavy reliance on a supply-oriented approach.

So far, periodic water shortages have not placed a substantial brake on the remarkable changes in Beijing's economy. Perhaps that will not always be so, as the loans against the future, in the form of unsustainable uses, come due, but it is also possible that Beijing's evolving water institutions are more resilient than they are given credit for.

Notes

1. Filial respects are due the honored ancestor of this chapter, Dixon, Fallon, and Nickum (1989). As usual in intergenerational matters, this descendant sees itself as more knowledgeable and up-to-date, with a broader (and quite different) perspective. At the very least, it is bigger. It has also benefited from the wise but nonbinding comments and suggestions of many more people, notably Maynard Hufschmidt, K. William Easter, Mei Xie, Yok-shiu Lee, Regina Gregory, Jennifer Turner, Parashar Malla, Jagadish Pokharel, Young-ho Chang, and Helen Takeuchi.

2. Because of its piedmont location and extensive area, average precipitation

in Beijing varies considerably between microclimates, ranging from under 500 mm on the leeward side of the mountains to over 700 mm in the Chaobai watershed (1956–1979 data from Wu and Hong 1988:4).

3. This includes the Huai River basin south of the Yellow River. Only in Beijing does the recharge rate exceed 500,000 m³/km²/year (Liu Changming and Wei 1989:69).

4. Beijing has actual precipitation data from the mid-1800s (Wu and Hong 1988:3). Up to 1979, there were 110 years of record (Bureau of Hydrology 1987:36).

5. Between 1986 and 1991, an average of 81 percent of mean annual precipitation occurred between June and September (ranging from 64 percent in 1990 to 94 percent in 1986) (source: various statistical yearbooks). Over the longer run, more than 85 percent has fallen during those 4 months (Wu and Hong 1988:3).

6. For the effect of "natural hazards" (*zaiqing*) on the fall harvest for the 1730–1915 period, see Zhang et al. 1992:29–30. On a scale of 0 (no disaster) to 8, only 2 years registered as high as a 6 (general disaster—probably implying a 30 percent crop loss).

7. Largely for these reasons, the Mongols selected the current location (Dadu/central Beijing) over the nearby site, now in the southwestern suburbs, of the central capital of the vanquished Jin Dynasty (Chen Cheng-siang [Chen Zhenxiang] 1981:113; Hong and Wang 1992:198–200).

8. A large-scale reservoir in China is defined as one with a storage capacity of over 100 million m³. A further distinction is made for large (I) reservoirs that can hold over 1,000 million m³. Only two of Beijing's 80-plus reservoirs, the Guanting and the Miyun, fit into the large (I) category. These two dominate Beijing's surface water supply capability. The Miyun, completed in 1959, has a capacity of 4.375 billion m³, and the Guanting (1954), 2.27 billion m³. Only the Haizi Reservoir (1960), which controls an isolated river system in Pinggu county, fits clearly into the large (II) category, with a capacity of 121 million m³, although the Huairou Reservoir (1958), now linked to the Miyun, is close at 98 million m³ (Liang 1989:51).

9. A similar expansion in groundwater use, linked to Green Revolution wheat, occurred in Punjab, India, at the same time.

10. This means that pumping was restricted to current-year discharge, without the buffer of over-year storage, not that continued abstraction was impossible. The transect shown in Figure 3.3 passes near the two industrial centers mentioned and hence may exaggerate the drawdown in areas off the line. Nonetheless, the drawdown has been dramatic and widespread throughout built-up Beijing.

11. In the mid-1980s, agricultural demands for water in the Shanxi "energy base" were calculated at 888 million m³ (83 percent of total demand) in a normal year (P = 50 percent), 772 million m³ (81 percent of total) in a dry year (P = 75 percent), and 660 million m³ (78 percent of total) in a very dry year (P = 90 percent) (Source: *Huabei diqu* 1985:18). The unwashed coal reference is from Nancy Yamaguchi, fellow, EWC, pers. com.

12. Unlike the Miyun, the Guanting Reservoir actually lies outside the borders of Beijing, just inside Zhangjiakou Prefecture.

13. For an early discussion of the merits and demerits of *nanshui beidiao*, with focus on environmental effects of the Eastern Route, drawing water from the Yangtze near Yangzhou, see Biswas et al., 1983. Recently, more serious consideration has been given to the Middle Route, which would take water from the Danjiangkou Reservoir on the Hubei-Henan border.

14. This section relies heavily on the joint study by the East-West Center and State Science and Technology Commission of China, 1988.

15. Another follow-up to previous studies has been to develop a macroeconomic water resource model based on a multiobjective goal programming methodology (North 1992). The model, developed over 3 years for a number of cities in north China and Ningbo, was completed in late 1992 under the Ministry of Water Resources, with assistance from the United Nations Development Programme.

References

Beijing Municipal Statistics Bureau, ed. July 1987. *Beijing Shi guomin jingji he shehui fazhan gaikuang 1981–1985* (Conditions of national economy and social development in Beijing Municipality, 1981–1985). Beijing: Zhongguo Tongji Chubanshe.

Biswas, Asit, Zuo Dakang, James E. Nickum, and Liu Changming, eds. 1983. *Long Distance Water Transfer in China*. Dublin: Tycooly International.

BJSTN86. September 1986. *Beijing Shi shehuijingjitongji nianjian 1986* (Beijing municipality socioeconomic statistics yearbook 1986). Beijing: Zhongguo Tongji Chubanshe.

BJSTN87. September 1987. *Beijing shehuijingjitongji nianjian 1987* (Beijing socioeconomic statistics yearbook 1987). Beijing: Zhongguo Tongji Chubanshe.

BJSTN88. August 1988. *Beijing shehuijingjitongji nianjian 1988* (Beijing socioeconomic statistics yearbook 1988). Beijing: Zhongguo Tongji Chubanshe.

BJSTN89. August 1989. *Beijing shehuijingjitongji nianjian 1989* (Beijing socioeconomic statistics yearbook 1989). Beijing: Zhongguo Tongji Chubanshe.

BJTN92. August 1992. *Beijing tongji nianjian 1992* (Beijing statistics yearbook 1992). Beijing: Zhongguo Tongji Chubanshe.

Bureau of Hydrology, Ministry of Water Resources and Electric Power. December 1987. *Zhongguo shui ziyuan pingjia* (China's water resource assessment). Beijing: Water Resources and Electric Power Press.

Chen Cheng-siang (Chen Zhenxiang). October 1981. *Zhongguo wenhua dili* (A cultural geography of China). Hong Kong: Joint Publishing Company (Sanlian).

Chen Jiaqi. January 1989. Huabei diqu shui ziyuan zhanlue wenti (Water resource strategies in north China). Presented to the Forum on Rational Development and Use of Water Resources held by the Earth Sciences Department of the Chinese Academy of Sciences, Beijing.

Chen Shenyi, Li Xianfa, and Long Qitai. March 1986. Beijing shuiziyuan gaikuang (Beijing's water resource conditions). Project report. Beijing Municipal Environmental Protection Scientific Research Institute, Beijing.

Chinese Research Team for Water Resources Policy and Management in Beijing-

Tianjin Region of China. August 1987. *Report on Water Resources Policy and Management for the Beijing-Tianjin Region of China.* For the Sino-Cooperative Research Project on Water Resources Policy and Management, Beijing.

Dixon, John A., Louise A. Fallon, and James E. Nickum. 1989. Water in Beijing: Conflict Resolution Over Space and Over Time. Paper prepared for the East-West Center and United Nations Centre for Regional Development Joint Workshop on Water Use Conflicts in Asian Metropolises, September 1989, Otsu, Japan.

EWC–SSTCC (East-West Center and State Science and Technology Commission of China). 1988. *Water Resources Policy and Management for the Beijing-Tianjin Region.* Honolulu, HI: Environment and Policy Institute, East-West Center.

Fallon, Louise A., and John A. Dixon. 1988. Vol. 2, "Scarcity without Shortage: Water Demand Management in the Beijing-Tianjin Region of China," in *Water for World Development.* Proceedings of the VI IWRA World Congress on Water Resources. Urbana: IWRA.

Han Huansheng and Li Xiangshu. September 1986. *"Liu wu" qijian woshi gongye jieshui qude xianzhu xiaoguo* (During the Sixth Five-Year Plan Our City's Industry Achieved Marked Results in Saving Water), in *Beijing shehuijingjitongji nianjian 1986.* Pp. 182–184. Beijing: Zhongguo Tongji Chubanshe.

Hong Shihua and Wang Jinru. 1992. "The Review and Prospect of the Water Resources Planning in Beijing," in proceedings of the *United Nations International Workshop on Water Resources Planning and Management in China,* Beijing, 2–6 April 1990. Pp. 197–209.

Hu Changzhi and Nie Shengyong. 1992. "Dachengbao xiezouqu" (A contracting concerto). *Zhongguo shuili,* 8:3–8.

Huabei diqu shui ziyuan gongxu yuce he jiejue di zhanlue cuoshi yanjiu (jieduan baogao) (Forecast of supply and demand for water resources in north China and a study of strategic measures for resolving [scarcity problems] [interim report]). December 1985. SSTCC Sixth Five-Year Plan Key Project Report No. 38-1-13. Hai River Commission of the Ministry of Water Resources and Electric Power, Beijing.

Hufschmidt, Maynard M., John A. Dixon, Louise A. Fallon, and Zhongping Zhu. August 1987. Water Management Policy Options for the Beijing-Tianjin Region of China. Draft report. Honolulu, HI: Environment and Policy Institute, East-West Center.

Johnson, Norman K., and Charles T. DuMars. 1989. "A Survey of the Evolution of Western Water Law in Response to Changing Economic and Public Interest Demands." *Natural Resources Journal* 29(2): 347–387.

Li Xianfa, Long Qitai, Nie Guisheng, and Wang Jie. 1987. "Better Water Resources Management—The Only Possible Solution to Quench Water Crisis," in proceedings of the *International Workshop on China's Water Environment Management,* Tongji University, Shanghai, May 1987. Pp. 155–182.

Liang Ruiju. 1989. "Management of Water Resources in Beijing," in *Proceedings of the IHP International Symposium-cum-Seminar on Integrated Water Management and Conservation in Urban Areas,* Nagoya. Pp. 47–55. Nagoya: International Hydrological Programme.

Liu Changming and Wei Zhongyi, eds. 1989. *Huabei pingyuan nongye shuiwen ji shui ziyuan* (Agricultural hydrology and water resources in the north China plain). Beijing: Kexue Chubanshe.

Milliken, J. Gordon, and Graham C. Taylor. 1981. *Metropolitan Water Management.* Washington, D.C.: American Geophysical Union.

Ministry of Water Resources and Electric Power, Planning and Design Institute. February 1989. *Zhongguo shui ziyuan liyong* (Water resource use in China). Beijing: Shuili Dianli Chubanshe.

Nickum, James E. March 1987. "Beijing's Rural Water Use." Working Paper for SSTCC–East-West Center Project on Water Resources Policy and Management for the Beijing-Tianjin Region, East-West Center.

Nickum, James E. In press. *Dam Lies and Other Statistics: Taking the Measure of Irrigation in China, 1931–1990* (tentative title). International Food Policy Research Institute, Washington, D.C.

Nie Guisheng, Su Wenhui, and Qin Datang. June 1986. *Beijing chengshi shui ziyuan xitong S.D. moxing moni fenxi* (Simulation analysis of Beijing's urban water resource system using an S.D. model). Report No. 14 for the Research Project on Water Resources Policy and Management in the Beijing-Tianjin Area, Beijing.

North, Ronald M. 1992. "Principles and Practices of Planning for Development and Management of Water Resources," in proceedings of the *United Nations International Workshop on Water Resources Planning and Management in China,* Beijing, 2–6 April 1990. Pp. 7–41.

Ren Guanzhao. 1991. *Zhongguo chengshi yongshui, jieshui wenti* (China's urban water use and conservation problems). *Shuili xuebao* 4:60–67.

Rural Socioeconomic Statistics Section, State Statistical Bureau. April 1989. *Zhongguo fenxian nongcun jingji tonji gaiyao 1980–1987* (Statistical essentials of the rural economy of China by county, 1980–1987). Beijing: Zhongguo Tongji Chubanshe.

Shabad, Theodore. 1972. *China's Changing Map.* Revised ed. New York and Washington: Praeger.

Water Resources and Hydropower Planning and Design Institute of the Ministry of Water Resources and Electric Power. February 1989. *Zhongguo shui ziyuan liyong* (Utilization of water resources in China). Beijing: Water Resources and Electric Power Press.

Wu Wengui and Hong Shihua, eds. June 1988. *Chengshi shui ziyuan pingjia ji kaifa liyong* (Urban water resource assessment and development for use). Nanjing: Haihe University Press.

Xie Mei. n.d. Decentralized Water Resources Management in Beijing, China. Mimeo.

Xie Mei, Nie Guisheng, and Jin Xianglan. 1991. "Application of an Input-Output Model to the Beijing Urban Water-Use System," in Karen R. Polenske and Chen Xikang, eds., *Chinese Economic Planning and Input-Output Analysis.* Pp. 239–253. Oxford: Oxford University Press.

Xie Mei, R. Rosso, G.L. Huang, and G.S. Nie. May 1989. "Application of Analytical Hierarchy Process to Water Resources Policy and Management in Beijing, China," in *Closing the Gap Between Theory and Practice.* IAHS Publication No. 180 (Proceedings of the Baltimore Symposium). Baltimore: IAHS.

Zhang Peiyuan, Ge Quansheng, Xu Kui, Zhang Jinrong, and Gong Gaofa. 1992. *18, 19 shiji woguo nongye qiushou zaiqing dengji xulie di fenxi* (Analysis of the sequence of agricultural summer harvest hazard grades in China during the 18th and 19th centuries), in Shi Yafeng et al., eds., *Zhongguo qihou yu haimian bianhua yanjiu jinzhan* (Studies on climatic and sea level changes in China), Vol. 2. Pp. 29–30. Beijing: China Ocean Press.

ZSN91. May 1992. *Zhongguo shuili nianjian 1991* (Almanac of China's water resources 1991). Beijing: Shuili Dianli Chubanshe.

ZTN90. August 1990. *Zhongguo tongji nianjian 1990* (China statistics yearbook 1990). Beijing: Zhongguo Tongji Chubanshe.

ZTN91. August 1991. *Zhongguo tongji nianjian 1991* (China statistics yearbook 1991). Beijing: Zhongguo Tongji Chubanshe.

4

Water Problems in Madras Metropolitan Region, India

R. Sakthivadivel and K. Venugopal

Madras, the capital of the state of Tamil Nadu and the fourth largest city of India, is a coastal city (see Figure 4.1). The areal extent of the city is 174 km². The land is flat, with coastal alluvium as top cover with a depth of not more than 50 m. The temperature ranges from 25° C to 37° C. The coldest and hottest months are January and May, respectively.

The area receives an average of 1,200 mm of rain per year, mostly during the southwest monsoon (June to September) and the northeast monsoon (October to December). The latter brings the heaviest rainfall. The northeast monsoon is intensified when depressions occur in the Bay of Bengal, creating storms and cyclones that cause flooding in Madras. Much of the rainfall flows into the sea unused. The maximum 1-day, 2-day, and 3-day rainfalls observed during 35 years (1951–1985) were 346 mm (1976), 588.8 mm (1985), and 726.6 mm (1985), respectively.

The population growth of Madras city from 1901 to 1981 is shown in Table 4.1, with phenomenal growth in the last two decades. The population of the Madras Metropolitan Area—which includes Madras city plus urban, semi-urban, and rural areas within a radius of 27 km—was 4,629,000 in 1981.

Industrial development in Madras city in the last decade was practically nil, unlike Calcutta, Bombay, and Delhi. Agricultural development also declined steadily in the Madras Metropolitan Area, as agricultural land, including the village tanks (ponds and reservoirs), was converted into housing plots. The Tamil Nadu Housing Board, part of the Tamil Nadu government, undertakes large-scale development of housing plots in and around Madras city, apart from numerous private agents.

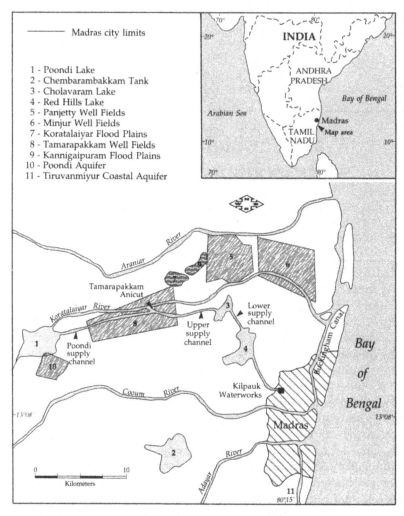

FIGURE 4.1 Water sources of Madras Metropolitan Area.

Among the principal cities in India, Madras has the lowest per capita consumption of drinking water in liters per day (see Table 4.2). The water problem in Madras city has become perennial. In order to fully understand the problem, we will examine the nature of surface water and groundwater resources in and around the Madras Metropolitan Area regarding quantity, quality, and projected demand.

TABLE 4.1   Population Growth of Madras City, 1901–1981

Year	Population	Year	Population
1901	541,167	1951	1,427,420
1911	555,800	1961	1,749,600
1921	578,550	1971	2,572,967
1931	713,394	1981	3,276,622
1941	865,334		

TABLE 4.2   Approximate Rate of Consumption of Water in Some Indian Towns

Town	Consumption (ℓ/capita/day)	Town	Consumption (ℓ/capita/day)
Agra	115	Hyderabad	205
Allahabad	125	Kanpur	205
Bangalore	90	Lucknow	90
Bhopal	100	Madras	70
Bombay	150	Patna	115
Calcutta	190	Poona	275
Delhi	160	Varanasi	115

Water Resources Potential in and Around Madras

Four rivers—Araniar, Koratalaiyar, Cooum, and Adayar—and the appurtenant reservoirs and tanks constitute the surface water resources for the Madras Metropolitan Area (see Figure 4.1).

Araniar River

Araniar is the northernmost river of Tamil Nadu state. It originates in Andhra state and flows through Tamil Nadu, joining the sea near Ponneri in Chingleput District. Of its total catchment area of 1,450 km², 700 km² lies in Andhra state and the rest in Tamil Nadu. The normal yield from the catchment area in Andhra state is used by that state. An average of 50 million m³ flows into Tamil Nadu state annually. The Surattapalle anicut (cross-regulator) just across the border diverts waters to fill a chain of tanks mostly in Tamil Nadu state. Two more anicuts, Annappanaicken-kuppam and Lakshmipuram, have been added downstream. The water is mostly used for irrigation. The level of land is such that it is possible to link this river with the Koratalaiyar River in the south by a channel to divert part of the flood surpluses.

Koratalaiyar River

The Koratalaiyar River originates from the surpluses of Kaveripauk tank located in North Arcot District. The surpluses of a number of tanks in the catchment downstream augment the flow. The Nageri area of this river, which drains a portion of Andhra state, joins the river a little upstream of Poondi Lake and brings substantial flow in the monsoon period. Of a total catchment area of 3,225 km², 790 km² lies in Andhra state. The annual yield of the basin is estimated at 580 million m³, of which about 57 million m³ is contributed by the catchment in Andhra state. The basin is mainly supplied by the erratic northeast monsoon. Tamarapakkam anicut across the river serves to divert water to Cholavaram and in turn to Red Hills Lake. Tamarapakkam has no surplus about 30 percent of the time, whereas the Red Hills Lake has surpluses about 50 percent of the time.

The present features of Cholavaram and Red Hills lakes are shown in Table 4.3. Although these two tanks were built for irrigation use, they now wholly serve as municipal water supply storages, with irrigation suspended since 1969.

Cooum River

The Cooum River starts as the surplus course of the Cooum tank in Tiruvallur taluk (subdistrict) and absorbs the surpluses of a number of tanks. The catchment area is about 290 km² studded with 140 tanks. A few minor irrigation tanks are fed through two anicuts across the river. The yield from this river is not of much significance for city water supply. It finds its way through the heart of Madras, serving as a sewage drain for the city.

Adayar River

The Adayar River is relatively short, only 42 km long, with a catchment area of 860 km². There are 450 minor irrigation tanks in its basin. The average annual yield of the Adayar has been computed as 140 million m³. Chembarambakkam tank, the largest of the old irrigation tanks of Tamil Nadu, lies in this basin. Its depth at full tank level (FTL) is 6.70 m with a capacity at FTL equal to 88.36 million m³ (with a water spread of 26 km²). Its catchment is about 358 km², with about 217 tanks adding surpluses into this tank. This is primarily an irrigation tank with a registered ayacut (command area) of 5,269 ha. Even with irrigation for its full command, one-third of its capacity can be spared. The tank has been serving as a year-to-year carryover reservoir.

TABLE 4.3 Features of Cholavaram and Red Hills Lakes

Lake	Full Tank Level (m)	Depth of Storage at the Deepest Point (m)	Capacity at Full Tank Level (million m³)
Cholavaram	19.65	5.45	22.97
Red Hills	14.69	5.80	80.65

The Buckingham Canal

This is a navigation canal parallel with and close to the seashore, excavated in the past century. It is a level bed canal linked to the sea at many points; it carries only seawater and its water level responds to the tidal variations of the bay. Its use for navigation has dwindled over the years because of poor maintenance.

Groundwater

The groundwater potential in and around Madras city has been identified with the help of United Nations Development Programme (UNDP) experts in three different phases: during 1966–1969, 1976–1978, and 1982–1985. In 1964, a United Nations mission visited Madras to investigate the feasibility of seawater conversion to supplement the water supply to Madras city. The mission realized and stressed the need for exploitation of groundwater potential instead. With the objective of evaluating the groundwater potential, the first phase of UNDP aid was given from 1966 to 1969. Three well fields—Minjur, Panjetty, and Tamarapakkam—in the Araniar-Koratalaiyar basin were identified. The safe yield was estimated to be 125 million ℓ/day. The second UNDP study from 1976 to 1978 was to cover all aspects of water supply and wastewater systems of the Madras Metropolitan Area. Groundwater potential in the coastal aquifer from south of Tiruvanmiyur to Kovalam was assessed as 1.15 million ℓ/day. In the third phase of the program, three new well fields—Poondi, Koratalaiyar, and Kannigaipuram—were identified in the Araniar-Koratalaiyar basin with a safe yield of 60 million ℓ/day.

Although the Palar River is popularly felt to be nothing but a dry sandy tract, detailed observations indicated a good supply of groundwater. Five aquifers are located along the river course at Athipatu, Pullambakkam, Pilapur, Manapakkam, and Vayalur. Groundwater potential of about 135 million ℓ/day is available. At present more than twenty municipalities and towns are drawing groundwater from the Palar basin.

Groundwater Quality Problems

The pH, electrical conductivity (EC), and ionic concentrations of the near-surface water vary significantly with place and time. The pH of the near-surface water ranges from slightly acidic (6.4) to alkaline (8.0). The EC of the water varies between 0.2 and 4.7 micro-mho per centimeter (μ-mho/cm) for different localities.

High concentrations of major ions near the Koratalaiyar River may be from recharge of saline backwater. The sulfate concentration is higher in the near-surface water than deep water and may be from recharge of effluent waters released by nearby industries and farms. The use of agricultural fertilizers has also resulted in higher potassium concentrations in shallow waters than in deep waters.

The pH of waters from deep wells is neutral, ranging from 6.8 to 7.2. The EC of deep water varies from 0.4 to 4.2 μ-mho/cm. The order of domination of cations and anions in deep water is similar to that of near-surface water. The concentration of all major ions is lower in the western part of the region and gradually increases toward the coast (i.e., toward the east). The variation of ion concentration suggests a steady increase of chloride in deep water from the dry to the wet season. The changes in values of ions are not appreciable when the water level is almost stable.

The chloride concentration is high in deep water, which may be from diffusion of seawater. In general, both deep and near-surface groundwaters are brackish to some degree in most of the study area. The increase in ratio of chloride to bicarbonate ions toward the sea demonstrates the entry of seawater into the aquifer, which is mainly because of overexploitation of groundwater by local and government agencies to meet the ever-growing demand of urban population and industries in and around Madras city.

The hardness of groundwater in Madras varies considerably. The shallow wells exhibit soft water, but the deep waters are generally hard. Alkalinity of groundwater is mainly caused by bicarbonates. In general, the groundwater of this aquifer is suitable for irrigation. In most of the area, however, the groundwater is not suitable for drinking because of high concentrations of Na+ and Cl– ions.

Generally, fluoride concentrations are very low, although moderate concentrations have appeared in Araniar-Koratalaiyar River water samples in selected locations. The European Community specifies a maximum admissible concentration of 1.5 mg/ℓ in drinking water to prevent fluorosis. The average concentration of fluoride in the sampled area is 0.733 mg/ℓ.

In the Araniar-Koratalaiyar basin at its confluence with the sea, the interface between salt- and freshwater advances landward. This intrusion, which was about 2 km inland in 1969, advanced up to 8 km by 1985.

Historical Development of the City's Water Supply

The city of Madras was earlier known as Chennaipatnam, and the domestic requirements then were completely met from backyard wells tapping the top aquifer. In 1872, Mr. Fracer, when he was special executive engineer of the Public Works Department (PWD), completed the Madras water supply scheme from surface water resources. Without a perennial river nearby and with the Cooum and Adayar rivers without a dependable yield and serving as local drainages only, the Koratalaiyar River was used for the city's water supply. This scheme was designed to provide a population of 0.47 million with 32 million ℓ/day by diverting the flows of Koratalaiyar River with an anicut at Tamarapakkam. The water is channeled to the Cholavaram irrigation tank through an upper supply channel. A lower supply channel links the Cholavaram tank to the Red Hills irrigation tank. From Red Hills Lake an open channel has been dug to convey water to the Kilpauk water treatment plant, where the distribution system starts. The water supply system to Madras city is shown schematically in Figure 4.2.

In 1914, Mr. Madley, corporation engineer of the Madras City Corporation, improved the system. An intake tower called Jones tower was constructed in the Red Hills Lake, and a closed conduit was constructed to convey water to Kilpauk water works.

The demand for water was growing steadily, and in the 1940s the diversion from these tank storages were found inadequate. Hence at about 30 km upstream of Tamarapakkam anicut, a new storage called Poondi Reservoir was created in 1944 on the Koratalaiyar River itself. A storage of 77.20 million m³ was provided, which raised the potential supply to 159 million ℓ/day.

The next effort to improve the supply came in the wake of acute scarcity experienced in the 1960s. The full-tank levels of both the Red Hills and Cholavaram tanks were raised by 0.61 and 1.22 m, respectively. Under the provisions of the Land Acquisition Act, 3,000 ha of wet lands (paddy fields) under the ayacut of the Red Hills and Cholavaram tanks were acquired by the government, at a cost of about 21 million rupees.

The Minjur, Panjetty, and Tamarapakkam well fields—identified during the first UNDP program, 1966–1969—supply water to industries (e.g., Madras Refineries, Madras Fertilizers) situated on the outskirts of Madras city. Some of this water is diverted to Madras city for supplementing the surface water supply whenever necessary.

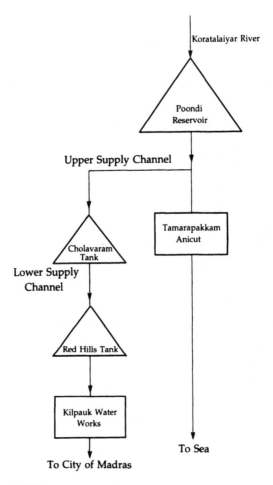

FIGURE 4.2 Water supply system to Madras city network.

In 1973, a separate head sluice was provided in the Poondi Reservoir. A lined channel was constructed to divert 2.85 m^3/sec directly to Tamarapakkam anicut to minimize seepage losses.

A part of the water from the Minjur well field has become saline from seawater intrusion. UNDP studies (1982–1985) suggested creating a water grid to connect the Araniar and Koratalaiyar rivers near the coast. The infiltrating water is expected to arrest further advancement of saline-water intrusion. The studies also suggested that extraction of groundwater in the area be controlled.

During drought, the owners of individual wells were selling water to lorries, and the lorries in turn sold the well water to buyers in the city. The government then enacted a law prohibiting private well owners from selling water and informed them they are entitled to draw their water requirements only. In 1987 the government of Tamil Nadu introduced licensing for bores and wells located in the eastern belt from Palavakkam to Injambakkam-Akkarai, as well as lorries transporting water to Madras. About 250 lorries and 60 wells and bores received licenses to supplement the water needs of Madras. The average cost of water available through lorries was Rs 150 per 12,000 liters. In September 1989 residents of Palavallam Panchayat noticed symptoms of saline-water intrusion resulting from excessive pumping and disallowed the lorries from taking water. The residents, through negotiations with the government, were successful in prohibiting the lorries from operating.

Present and Future Water Demand

Present Water Demand

The 1981 census revealed population of 3.3 million in Madras city, 1 million in the urban areas bordering the city, and 0.34 million in the rural areas, totaling 4.64 million in the Madras Metropolitan Area. However, knowledge of population figures alone is not enough to estimate demand for water. Upper income groups require more water than low-income groups and people who live in squatter settlements or "slums." This is because people in upper income groups are provided with house service connections, whereas "slum" dwellers are served by street taps. Low-income groups include both types of users. Besides the economic stratification of population, per capita consumption is influenced by factors such as climatic conditions, habits of people, continuous or intermittent water supply, availability of water from private sources, and pressure in the pipeline. Also, water use increases with availability of water-carried sewerage services. For conditions in India, the basic water requirements for domestic consumption is about 107 ℓ/capita/day. Indian standards for an urban area is 135 ℓ/capita/day. Since Madras city is blessed with a shallow water table, and most of the houses are provided with wells to service nondrinking requirements, a per capita supply of 100 ℓ/capita/day on average is considered reasonable to meet the requirements in prime residential areas.

The per capita requirement for commercial areas is also 100 ℓ/capita/day. In slum areas where residents are served by standpipes, water is typically handcarried, and the actual consumption is limited to minimum requirements. Random sampling of public standpipes indicates the per capita consumption for slum dwellers is about 40 ℓ/capita/day.

TABLE 4.4   Water Consumption for Different Types of Buildings

Type of Building	Consumption (ℓ/head/day)
Factories where no bathrooms are required to be provided	30
Factories where bathrooms are required to be provided	45
Hospitals, including laundry (per bed)	
Not exceeding 100	340
Exceeding 100	455
Nurses' homes and medical quarters	135
Hostels	135
Hotels (per bed)	180
Offices	45
Cinemas, concert halls, and theaters (per seat)	15
School	
Day	45
Boarding	135

Water use in mixed residential areas (with a mixture of prime residential and slum population) averages 80 ℓ/capita/day. Considering that the present population of 4 million for Madras city consists of 1.86 million living in prime residential areas, 0.22 million in commercial areas, 1.1 million in mixed residential areas, and the remaining 0.82 million in slum areas, the domestic water demand for each category comes to 186, 22, 88, and 32.8 million ℓ/day, respectively. The total demand is about 330 million ℓ/day.

A physical count was conducted in 1979 to assess water use for various institutions and for different types of nondomestic buildings. Institutional water use varied around a mean of approximately 20,000 ℓ/ha/day. The per capita consumption of water for different types of buildings used in the computations are shown in Table 4.4.

The water requirements for industries can be classified under two separate categories: inside the city and outside the city. The requirements inside the city are generally met from the piped water supply system, but for most of the bulk consumers like railways or the Port Trust, a special pipeline has been provided so that the industrial requirements are not affected because of domestic demand and vice versa. The present withdrawal for these large industries inside the city is 7 million ℓ/day (Table 4.5). The requirements of industries outside the city are met from the groundwater grid. The present consumption for industrial use outside the city amounts to about 76 million ℓ/day (Table 4.6). Thus, the total demand for industries inside and outside the city is 83 million ℓ/day, and the total of all requirements amounts to 413 million ℓ/day.

TABLE 4.5 Present Industrial Requirements Inside Madras City

Name of Industry	Quantity Actually Supplied (million ℓ/day)
Madras Port Trust	2.5
Southern Railway	1.8
Basin Bridge	0.6
B & C Mills	1.0
Others	1.1
Total	7.0

The preceding figures do not account for leakage. Leakage studies carried out in the Madras city distribution system in 1974 and 1979 indicated losses of about 12 percent, as the operating pressure is low and availability is always less than demand. Leakage tends to be higher with an increased rate of water supply, which will result in higher pressure in the system; hence, a 15 percent loss is assumed. Accounting for losses, the present demand is about 485 million ℓ/day.

TABLE 4.6 Present Industrial Requirements Outside Madras City

Name of Industry	Quantity Actually Supplied (million ℓ/day)
1 Madras Refineries Limited	16.10
2 Southern Railway (Ennore)	4.60
3 Ennore Thermal Power Station	11.70
4 Madras Fertilizers Limited	19.60
5 Indian Organic Chemicals Limited	1.70
6 Central Reserve Police Force	1.40
7 Eothari (Madras) Limited	1.30
8 Linear Alkyl Benzyme	6.80
9 Others	2.40
Subtotal	65.60
10 Dunlop India Limited	1.80
11[a] Shaw Wallace & Company Limited	4.00
12[a] Compound Fertilizer Factory	3.30
13[a] Others	1.70
Subtotal	10.80
Total	76.40

[a]Supplied by own/other agencies.

TABLE 4.7   Madras Metropolitan Area: Population Projection (in thousands)

Area	1981	1991[a]	2011[b]
Madras city	3,277	4,330	7,300
Urban centers surrounding the city	1,012	1,340	2,700
Rural	340	370	600
Total	4,629	6,040	10,600

[a]As per Madras Metropolitan Development Authority.
[b]Estimate.

Water Demand for 2011

Population projections for 1991 and 2011 are compared with that of 1981 in Table 4.7. See Table 4.8 for the estimated distribution of population in squatter settlements, low income, and upper income categories in the Madras Metropolitan Area for the year 2011. The projected population of 7.3 million for Madras city is expected to consist of prime residential, 3.4 million; commercial, 0.4 million; mixed residential, 2.0 million; and slum, 1.5 million.

Using the per capita figures mentioned previously for each land use type, the domestic water demand in Madras city will be prime residential, 340 million ℓ/day; commercial, 40 million ℓ/day; mixed residential, 160 million ℓ/day; and slum, 60 million ℓ/day.

Planners anticipate an institutional land use of 2,000 ha by 2011, and at 20,000 ℓ/ha/day, the water demand would be 40 million ℓ/day. To account for public purposes (e.g., maintenance of baths, sewer cleaning, road watering), the institutional requirements would double, and water demand would be 80 million ℓ/day.

The future demand for industries inside the city limit is estimated to

TABLE 4.8   Madras Metropolitan Area Population Characteristics, 2011 (in thousands)

Population Type	Urban		Rural	MMA (total)	Percent
	City	Centers			
Squatter settlements	1,460	540	90	2,090	20
Low income	2,190	1,620	450	4,260	40
Others	3,650	540	60	4,250	40
Total	7,300	2,700	600	10,600	100

MMA = Madras Metropolitan Area.

be 14 million ℓ/day, and for industries outside the city limit, 152 million ℓ/day (the demand was simply doubled for both cases).

In summary, water demand for the year 2011 for Madras city would be about 1,000 million ℓ/day: residential, 600 million ℓ/day; institutional, 80 million ℓ/day; industrial, 166 million ℓ/day; plus 15 percent leakage, 127 million ℓ/day.

Measures Adopted/Proposed to Meet the Demand

At present the Madras city water supply is being handled by the Madras Metropolitan Water Supply and Sewerage Board (MMWSSB), which was formed in 1978. The organization is mainly responsible for distributing water to the entire metropolitan area. The schemes for developing new water sources lie with both the Tamil Nadu Public Works Department (PWD) and the MMWSSB. Groundwater development is handled by the MMWSSB and the groundwater division of PWD, whereas surface water development is handled mainly by the PWD itself. The development of urban areas around the city is handled by the Madras Metropolitan Development Authority (MMDA). The MMDA is mainly responsible for approving development of plots and plans for certain types of buildings. The Tamil Nadu government is planning and developing a few proposals to meet the water needs of the Madras Metropolitan Area.

Krishna Water Supply Project

The Krishna Water Supply Scheme envisages an annual supply of 340 million m³ of water at the Tamil Nadu border with Andhra Pradesh. The Krishna River, which is about 1,400 km long, originates in the western Ghats in Maharashtra state and flows through Karnataka and Andhra Pradesh. Its flow is mainly derived from the southwest monsoon during the months of June to September. The annual flow of the river at 75 percent dependability has been assessed as 58,340 million m³, and much of it continues to flow waste during floods. In the past 16 years, annual discharge to the sea averaged 22,650 million m³. The project scheme is simple but massive. Plans call for drawing water from the Srisailam Reservoir when the Krishna River floods from July to October and for conveying water to the Somasila Reservoir on Pennar through an open canal. A reservoir will be built at Kandaleru, linking it to the Somasila Reservoir by another canal. From Kandaleru a canal will be constructed to the Poondi Reservoir, crossing the state of Andhra Pradesh to Tamil Nadu.

In order to increase the storage capacity of Poondi Reservoir by 20 million m³, the full reservoir level will be raised by 0.6 m. Upstream of Poondi Reservoir, a storage capacity of 33.40 million m³ is proposed for the Ramanjeri Reservoir. Downstream of Poondi Reservoir, 28.30 million m³ of storage capacity is proposed for the Thirukkondalam Reservoir. Besides the Poondi Reservoir, full reservoir levels of Red Hills and Chembarambakkam tanks will also be raised. By raising the levels by 0.6 m, an additional storage of 12.75 and 14.85 million m³ is expected in the tanks of Red Hills and Chembarambakkam, respectively. Poondi Reservoir will be connected to Chembarambakkam through a link canal and from it to Red Hills tank by a feeder canal.

The ayacut area under Chembarambakkam tank is being slowly converted into housing plots. The Tamil Nadu government is waiting for a suitable time to acquire the irrigation rights of the ayacut existing under the Chembarambakkam tank. About Rs 130 million has been allocated to purchase these rights.

The government initially expects 400 million ℓ/day of water to reach Madras by 1993–1994. The Tamil Nadu government has drawn a scheme for about Rs 5.5 billion and is seeking aid from the World Bank.

Groundwater Development

UNDP studies conducted during 1982–1985 indicated that a groundwater potential of about 135 million ℓ/day is available in Palar Basin in five aquifers. Present withdrawal, however, is only one-third of the groundwater potential. Table 4.9 indicates the location of the aquifers, groundwater potential, and other details. A study by MMWSSB under the UNDP II Project started in 1987 confirmed the availability of about 25–45 million ℓ/day of unused potential in aquifers at Manapakkam and Vayalur.

Oggian Madagu

Oggian Madagu is a small swamp area, situated south of Madras. Its length varies from 5 km immediately after the monsoon to 1 km in summer. The water dries up by February and March of every year and is replenished from the monsoon rains. The inflow into Oggian Madagu is rainwater. However, because it is connected to the Buckingham Canal, which has a number of openings to the sea, the water is saline and also high in total solids.

Although the present volume of Oggian Madagu is small, it is attractive in that it lies near Madras and can be developed as a potential water supply source. The storage in Oggian Madagu can also be used to recharge the Southern Coastal Aquifer. A number of tasks must be done to develop Oggian Madagu: it must be deepened and provided with regular bunds; the section of Buckingham Canal passing through Oggian Madagu must have locks for the passage of boats and to prevent saltwater from mixing with freshwater in Oggian Madagu. After these tasks

TABLE 4.9   Details of Groundwater Potential in Palar Basin

Location	Distance from Madras (km)	Groundwater Potential (million ℓ/day)	Present Withdrawal (million ℓ/day)
Panapakkam Athipattu	94	23	4.5
Pullambakkam	61	26	11.5
Pilappur	55	9	4.5
Manapakkam	65	18	—
Vayalur	70	59	25.0
Total	345	135	45.5

are completed, freshwater collected in the Madagu for two or three seasons is expected to flush out all the saline water. The swamp can then be maintained as a freshwater reservoir. The feasibility of this project is being explored.

Other Measures

Although the Tamil Nadu government is eager to give priority to domestic water demands as opposed to water for the industrial and agricultural sectors, so far no effective legal or pricing mechanisms have been introduced to that effect. The metropolitan development itself—by private entrepreneurs and by state government undertakings like the Tamil Nadu Housing Board, Slum Clearance Board, and Harijan Housing Corporation—has led to large increases in land value for housing and to a general desire among the people to own a house for each family. Agriculturists, who receive meager income from agriculture, find it more economical to sell their land and buy agricultural land elsewhere or resort to other forms of investment. This tendency has by itself reduced the area under irrigation steadily, favoring the thinking of government. During water shortages, water is supplied on alternate days or once in 3 days, leading to temporary closure of water-intensive industries. Educational institutions accelerate their courses and close down for a prolonged period. The MMWSSB and private agencies commonly supply water by lorries to hotels and marriage halls. MMWSSB lorries are also used to fill public tanks in certain areas. MMWSSB maintains about 525 lorries and trucks for supplying water to households during acute shortages.

Policy Options to Manage City Water Problems

Managing the city's water problems with high water demand and low water supply has become a way of life. The problem needs a systematic approach for using both surface water and groundwater from every available source, including necessary hardware components and sophisticated software strategies. "Crisis management" are the key words in dealing with water problems.

As mentioned earlier, groundwater resources are used to supplement surface water supplies. Investigations are being conducted to estimate the safe yield from existing well fields and to locate new and promising well fields. Integrated management of surface water and groundwater is being studied to minimize evaporation losses, as the previous work for reducing evaporation did not show promising results.

To minimize flood flows to the sea, a link canal to connect the Araniar River to the Koratalaiyar River, and check dams all along the river course, are planned to augment groundwater storage. Diverting the groundwater flow, which otherwise will flow into the sea, is also proposed. Rehabilitation of tanks in the area is also being considered. The storages of the existing lakes (i.e., Poondi, Cholavaram, and Red Hills) are being enhanced by increasing the full reservoir levels by 0.6 m.

The requirement that irrigators pay for the right to divert water from tanks is in the discussion stage.

MMWSSB is encouraging large-scale industries to reuse treated sewerage water so that freshwater can be made available for domestic purposes. To discourage migration of people to the city, the Madras Metropolitan Development Authority (MMDA) is encouraging development of plots and flats with basic amenities in the surrounding urban agglomeration. Private agencies are also encouraged to develop house sites following the land use plans prepared by MMDA. Improvements in the transportation network are also being implemented to encourage people to move to the outskirts of the city. Areas where groundwater availability is poor are zoned for industrial purposes.

Also, the central market, Kothaval Chavadi, is being relocated to Koyambedu on the outskirts of the city.

Conclusion

In essence the water shortage problem of Madras city is acute. Unless proper measures are taken, the city will inevitably suffer. All possible alternatives should be listed, and pertinent data collected for each alternative to conduct technical and economic feasibility studies. The time and money spent toward planning should not be construed as wasteful.

The dependable yield from surface water sources for meeting domestic water needs and the adequacy of existing storage should be investigated. Diverting water from other basins should be pursued to meet expected future water demand.

The development and use of groundwater should be carefully monitored to keep the withdrawal within safe limits and also to prevent saltwater intrusion. To maintain groundwater quality, the Buckingham Canal should be cleaned and the effluent from industries should be treated to a specified standard. Groundwater recharge should be taken up in large measure.

Rules and regulations enacted should be such that they will be followed without exception. Awareness should be created among the public to use water efficiently. "Conserve water" should be a key slogan in the minds of the people.

5

Urban Water Management in Metropolitan Manila, Philippines

Francisco P. Fellizar, Jr.

The concern for urban water management and conservation is timely and relevant against the backdrop of burgeoning population in most urbanized and urbanizing areas in the Asian region. Inevitably, these centers must cope with managing demands and ensuring a steady supply to meet increased demands for safe and usable water. This requires a deeper appreciation and understanding of the issues and factors that influence the water supply-and-demand situation within a given societal context. It is imperative, for instance, to seriously consider culture, resources, policies and institutions, development priorities, and the environment, among other things, when planning for sustainable water resource management programs.

Like other urban areas in Asia, Metropolitan Manila has its share of problems arising from increasing demand for water and the burdens of providing for this demand.

This paper presents some of the issues and corresponding measures currently undertaken, as well as actions taken in the past relevant to water supply management and conservation. Likewise, practical and policy-related concerns are raised.

Philippine Water Resources: Facts and Figures

The Philippines consists of about 7,000 islands and islets with an aggregate area of approximately 300,000 km² (NWRC 1976:1). The tropical Philippine climate shows more variation in rainfall than in temperature. Hence, the climatic classifications are based mainly on the occurrence of wet and dry seasons.

The precipitation has a wide range of variability. For instance, the highest intensity rainfall was 9,006 mm in Baguio City in 1910, and the lowest was 94.2 mm in Vigan in 1948. Average rainfall per year is 2,360 mm.

As for water resources, the Philippines is fortunate to have been bountifully endowed. It has about 421 principal rivers and 59 natural lakes, aside from numerous streams. Moreover, four major groundwater reservoirs, when combined with other smaller reservoirs, would have an aggregate area of about 50,000 km². The total amount of groundwater storage is estimated at 260,000 million m³ and the rate of net groundwater inflow at 33,000 million m³/year (NWRC 1980:7–12). These resources are considered more than sufficient to meet the water requirements of the country well beyond the year 2000. In fact, the amount of surface water available 90 percent of the time alone would be more than twice as much as total demand projected for the year 2000. Despite the apparent abundance of water resources in the country, area-specific scarcity exists because of the natural distribution of supply sources. Thus, comprehensive planning and development of water resources remain an important undertaking of the government. This concern is further exacerbated by the increasing demand for water for a variety of uses by the growing population.

Realizing the vital role of water resources management and conservation to national development, the government promulgated a National Water Code as early as 1976, to provide a comprehensive and rational basis for integrated and multipurpose management of water resources. This document, more popularly known as the Water Code of the Philippines of 1976, has the following objectives (NWRC 1982:3):

- To establish the basic principles and framework relating to the appropriation, control, and conservation of water resources to achieve the optimum development and rational utilization of these resources;
- To define the extent of the rights and obligations of water users and owners, including the protection and regulation of such rights;
- To adopt a basic law governing the ownership, appropriation, utilization, exploitation, development, conservation, and protection of water resources and rights to land related thereto; and
- To identify the administrative agencies that will enforce this Code.

The Code has the following underlying principles (NWRC 1982:3):

- All waters belong to the State.

- All waters that belong to the State cannot be the subject of acquisitive prescription.
- The State may allow the use or development of waters by administrative concession.
- The utilization, exploitation, development, conservation, and protection of water resources shall be subject to the control and regulation of the government through the National Water Resources Council (now known as the National Water Resources Board). The Board is composed of department heads (formerly ministries) and line agencies particularly concerned with water resources.
- Preference in the use and development of waters shall consider current usages and be responsive to the changing needs of the country.

According to the stated objectives and principles of the Code, appropriation of water through the use of water permits is evidence of the water rights granted by the government. The measure and limit of appropriation of waters shall be beneficial use, which refers to water use in the right amount during the period that the water is needed for producing benefits for which the water is appropriated. The standards of beneficial use shall be prescribed by the National Water Resources Board. Between two and more appropriators of water from the same source of supply, priority in time of appropriation shall give the better right, except that in times of emergency the use of water for domestic and municipal purposes shall have a better right over all other uses. When water shortage is recurrent and the appropriator for municipal use has a lower priority in time of appropriation, then it shall be his/her duty to find an alternative source of supply in accordance with conditions prescribed by the Board (NWRC 1982:9).

To achieve management and development of water resources consistent with the principles of optimum utilization, conservation, and protection for present and future needs, the Board coordinates the activities of a number of governmental and private agencies that perform specific functions in water resources development (e.g., the National Irrigation Administration, the National Power Corporation, the Local Water Utilities Administration, and the Manila Metropolitan Waterworks and Sewerage System). The Board also makes appropriate recommendations to the President of the Republic through the National Development Authority on matters relating to water resources.

Within this context, the issues and challenges related to water resources management and conservation in Metropolitan Manila shall be presented in the succeeding sections.

Metropolitan Manila: Geographical Description

Location

Metropolitan Manila is located along Manila Bay (see Figure 5.1). Created by Presidential Decree no. 824 in 1975, it is composed of four cities and thirteen municipalities. As the capital of the Philippines, Manila is the commercial, industrial, cultural, and administrative center of the country. It covers less than 3 percent of the total land area (636 km²) but accounts for 14 percent of the population, 50 percent of all large enterprises, and 33 percent of all manufacturing units.

Population

From the 1982 estimate of about 5.92 million, the population was expected to reach 8.89 million in 1990 and 11.0 million by the year 2000 (NEDA 1982). Annual population growth ranges from 3.4 to 3.8 percent. The population density of 1984 was estimated at 7,774 persons/km², with the city of Manila having the highest density of 39,977 persons/km² (Palencia 1984: 70).

Water Sources

The 90-year average rainfall is estimated to be 2,070 mm (82 inches) per year. In a 1971 study on water resources in the Manila Bay region, it was estimated that there exists a total potential water supply of 30 billion m³—an annual rainfall water supply of 21.5 billion m³ and an additional 8.6 billion m³ from groundwater resources (MWSS 1986). Thus, in Metropolitan Manila, the problem of access to water is not from shortage but from underinvestment and managerial problems (IEP 1971).

The Water System in Metropolitan Manila: A Historical Background

Carriedo Waterworks (1882–1919)

Before the "discovery" of the Philippine Islands in 1521, the residents of Manila drank rainwater from surface wells. Water was carried in jars called *tinajas* and transported downstream in small boats for distribution.

In 1690, a Franciscan Friar, dissatisfied with this method, developed a spring located in San Juan del Monte, a municipality outside Manila, and constructed an open aqueduct and a small navigable canal to make spring water accessible to Manila residents. In 1743, a retired Spanish Captain General, Don Francisco Carriedo y Peredo, bequeathed his entire fortune of ₱10,000 to Manila as the nucleus of a fund for public water

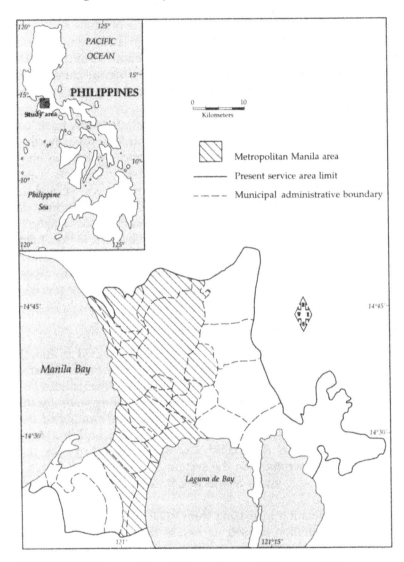

FIGURE 5.1 Metropolitan Waterworks and Sewerage System area (*Source* of base map: PICOREM 1980).

supply for the city. This donation, with accumulated interest, was used for construction of the Carriedo Waterworks.

The Carriedo Waterworks, completed in 1882 at a cost of ₱745,509, was the first community water system in Manila and one of the first properly designed water supply systems in the Far East. Under the system, water was pumped from the Marikina River at Santolan into "El

Deposito" in San Juan, an underground reservoir of adobe and tiles, from where it flowed by gravity to the city through a 650-mm transmission main. The capacity of the system then was about 15 million ℓ/day. In 1898, shortly after the American occupation, two more pumps were installed at the Santolan Plant, which increased the system's capacity to about 30 million ℓ/day.

However, water from this source became polluted with waste coming from several towns along the banks of the Marikina River north of the pumping plant. A decision was made to construct a gravity system with an intake upstream from the town of Montalban, where the watershed could be protected from contamination. The project was started in 1902 and completed in 1909. It included construction of a masonry dam at Wawa (Montalban) and 24 km of steel pipe and concrete aqueduct leading to a 212-million liter reservoir at a high elevation in San Juan. The distribution system in Manila was also improved and expanded. The entire project cost was about ₱6.0 million and had a maximum capacity of 87 million ℓ/d. The old pumping plant and reservoir were placed in reserve when this new system began operation.

Metropolitan Water District (1919–1955)

In 1919, the Metropolitan Water District was created to handle the problems of water supply for Manila and its suburbs. Its service area was expanded to include fourteen adjoining cities and municipalities.

The Angat-Novaliches System, storing and diverting water for the first time from the Angat River, was begun in 1925 and completed in 1939. The capacity of this system, based on the Ipo and La Mesa dams, increased from an initial 300,000 m³/day up to 380,000 m³/day in 1953, despite the havoc wrought on the system by World War II. Since funds were inadequate, only minor repairs and improvements were undertaken after the war.

With the population increasing from 913,000 in 1939 to 1.6 million in 1948, and to 2.5 million in 1960, as well as the burgeoning of industries and subdivisions, water demand increased tremendously. The Angat-Novaliches System became inadequate to meet metropolitan needs.

National Waterworks and Sewerage Authority (1955–1971)

On 18 June 1955, the National Waterworks and Sewerage Authority (NWSA) was created to take over the functions of the Metropolitan Water District and to centralize and consolidate under its control, direction, and general supervision all waterworks and sewerage systems in the country.

In 1956, a second aqueduct from Novaliches to Balara was constructed to convey more water to the Balara Treatment Plant. In 1957, the crest of the Ipo Dam was raised by 2.0 m to increase the capacity by 57 million ℓ/day, for a total of 760 million ℓ/day. The crest of La Mesa Dam was also raised by 2.0 m to provide additional storage.

In 1962, a small dam was constructed on the Alat River and an aqueduct built from the Alat River to the Novaliches Reservoir, thereby increasing the combined yield of the Alat-Novaliches supply to about 164 million ℓ/day.

In 1962 and 1967, two temporary pumping stations were constructed in the Marikina River to pump water to the Balara Treatment Plant during high river flows, thereby maintaining higher water levels in Novaliches Reservoir. Each station has a pumping capacity of 189 million ℓ/day. Over the year, wells were constructed to serve local communities.

The interim program of improvement (later called Manila Water Supply Project I) was begun in 1964 after intensive studies were made in conjunction with the construction of the Angat Multipurpose Dam in San Lorenzo (completed in 1967 with NWSA sharing ₱21.5 million) wherein a total draw of 1,895 million ℓ/day was allocated for water supply. Financing was obtained through a loan of $20.2 million from the World Bank (International Bank for Reconstruction and Development) and local funds.

Projects undertaken starting in 1965 consisted of a new Ipo intake structure; a second Ipo-Bicti tunnel of 1,137 million ℓ/day capacity; a third aqueduct from Bicti to Novaliches of 758 million ℓ/day capacity; a second Balara filtration plant; ten pumping stations and reservoirs located at strategic places in Metropolitan Manila; and about 100 km of distribution mains, ranging from 300 to 1,500 mm in diameter.

The interim program was completed in 1972 at a total cost of ₱477 million and increased the delivery of water to a total of 1,137 million ℓ/day. Nevertheless, water demand grew faster than supply, resulting in water shortages year after year.

The Current Water Supply System

The Metropolitan Waterworks and Sewerage System (MWSS), a government-owned and -controlled corporation, was created on 19 June 1971, superseding the National Waterworks and Sewerage Authority to continue providing potable water supply and sewerage system services for Metropolitan Manila and its environs. Its service area initially covered five cities and twenty-three municipalities totaling about 148,700 ha (MWSS 1989a:1–6). The coverage was expanded subsequently to include an additional area of Rizal Province.

The population of the service area was 8.287 million in 1988 and is growing at a rate of approximately 3 percent per year. Population in the area is expected to reach over 11.0 million in the year 2000.

MWSS Organization and Finance

The MWSS Board of Trustees, composed of seven members, exercise the corporate powers and functions of MWSS. Members include the Department of Public Works and Highways (DPWH) secretary as ex-officio chairperson, an administrative vice-chairperson, and five members. The administrative vice-chairperson and four members are appointed by the President of the Philippines. The remaining member is the government corporate counsel, an ex-officio member, who acts as the legal advisor of the board. The board formulates policies, approves budgets, recommends tariff adjustments, and issues necessary regulations for conducting MWSS business. A management staff takes care of the day-to-day supervision of the corporation's activities. This group is headed by the administrator, assisted by a senior deputy administrator plus six deputy administrators, who respectively take charge of engineering, construction management, operations, customer service, finance, and administration.

The MWSS water supply projects are financed by a combination of government grants as equity to the corporation, internally generated funds, and loans from international lending institutions. The system's initial capitalization was ₱3.0 billion. This increased to ₱8.0 billion in 1985, of which ₱2.6 billion was paid up as of December 1985. Corporate revenues are used to finance operating expenses, loan repayments, and tax liabilities. Government equity is used as counterpart funds for foreign-assisted projects, for debt-servicing, and to defray a portion of MWSS operating expenses. The MWSS had to rely on international financing institutions to cover approximately 50 to 70 percent of its major investments. As of year-end 1985, MWSS had outstanding obligations of $112.6 million to the Asian Development Bank, $112.9 million to the World Bank, and ₱54.0 million to the Philippine National Bank. The MWSS has a total net worth of ₱8.53 billion (see Table 5.1 for a financial profile of MWSS).

MWSS is expected to achieve its goal of becoming a financially viable organization. Water and sewerage tariffs are prescribed to provide for sufficient revenues to cover all costs of operations and maintenance, depreciation, and debt-servicing. However, regulations governing public utilities limit net profits to 12 percent of total assets. The upward trend of MWSS operating revenues and net income reflects the increases in tariff levels, production capacity, and service coverage. Because of increasing demand for water and the concomitant need for investment, plus repaying foreign loans, there is indeed the imperative of enhancing the financial capability of the MWSS.

TABLE 5.1   Financial Profile of MWSS (million pesos)

	1981	1982	1983	1984	1985
Sources of funds					
Operations	244.01	292.94	313.77	456.55	746.01
Long-term					
borrowings	342.24	438.85	1,267.28	1,711.78	619.15
Additional					
capitalization	330.00	285.00	366.00	300.00	600.00
Various sources	11.95	26.70	74.09	44.66	42.86
Financial condition					
Assets	6,099.70	7,564.97	9,341.62	11,928.60	13,441.36
Liabilities	881.20	1,368.64	2,751.46	4,655.33	4,914.43
Net worth	5,218.50	6,196.33	6,590.16	7,273.27	8,526.93
Results of operations					
Operating revenues	381.56	462.31	523.92	706.37	1,133.58
Other income	62.53	58.80	38.71	93.05	73.70
Expenses	251.79	287.58	349.68	453.46	590.59
Net income	192.30	233.53	212.95	345.96	616.69

Source: MWSS 1986.

The Water Supply Situation

Table 5.2 shows water supplied by MWSS between 1984 and 1988. The main source of raw water is the Angat/Ipo Stream Reservoir System, which is approximately 50 km north of Manila. Other sources include the Alat-Novaliches Reservoir and a small number of wells.

Figure 5.2 shows the MWSS water supply sources and distribution system. Water from the Angat Reservoir is released and diverted to the Ipo Dam, where it converges with water from the Ipo River. The water is then conveyed by tunnel and pipe aqueduct to the Novaliches Reservoir, where it is temporarily stored before it goes to two treatment plants— Balara in Quezon City (capacity 1.6 million m³/day) and La Mesa in

TABLE 5.2   MWSS Water Supplied (million m³)

	Source of Water			Groundwater
Year	Surface	Ground	Total	(%)
1984	642.24	25.56	667.80	3.83
1985	757.37	29.45	786.82	3.74
1986	874.07	30.43	904.50	3.36
1987	834.75	27.87	862.62	3.23
1988	849.34	29.48	878.82	3.35

Source: MWSS 1989a.

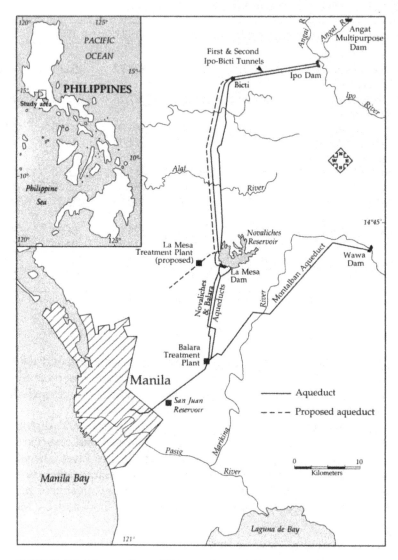

FIGURE 5.2 Metropolitan Waterworks and Sewerage System water supply sources and distribution system.

Novaliches (capacity 1.5 million m³/day). From these two plants, treated water enters the central distribution system. Water is then supplied to customers through metered service connections, which as of 1985 totaled 437,000. There are also about 1,364 public standpipes, mainly in blighted areas (MWSS 1986).

TABLE 5.3   Population Served by MWSS

Year	Population Under MWSS Jurisdiction (million)	Population Actually Served (million)	(%)
1984	7.364	2.832	38.5
1985	7.597	3.288	43.3
1986	7.826	3.810	48.7
1987	8.053	4.197	52.1
1988	8.287	4.492	54.2

Source: MWSS 1989a.

Table 5.3 shows the population served by MWSS from 1984 to 1988. There was a steady increase in the percentage of population served, although in small yearly increments. Considering the rate of population growth in the region, these increments are positive indications of MWSS efforts to cater to more people. But a significant portion of the population suffers from insufficient drinking water, resulting from absence of water mains in some areas; illegal connections; unrepaired leakage in the pipe system; low pressure in the secondary system; drying up of underground wells; and improper and increasing use of booster pumps (verified through observations, selected interviews, and various MWSS reports).

Since most residents own their wells, private groundwater withdrawals are quite high. One document indicated that from 1980 to 1981, groundwater contributed about 40 percent of the supplies for the Greater Metropolitan Manila (interviews with MWSS staff, October 1989). The aquifers in the metropolitan area are grossly overexploited, which has caused lowering of groundwater levels by as much as 4 to 10 m/year (MWSS 1989b:32). Lately, seawater intrusion from Manila Bay has increased significantly. Thus, the cost of pumping groundwater in most areas has become financially prohibitive. Most affected are the subdivision owners and developers situated in areas with critical water conditions. These problems exert pressure on the MWSS to expand its effective service area and to ensure alternative supply sources, which undoubtedly would mean a huge financial investment.

Nonrevenue Water (NRW)

Much of the water supplied by MWSS is nonrevenue water (i.e., water produced and delivered without corresponding revenue). From 1973 to 1987, the percentage of NRW registered a remarkable increase, particularly from 1978 to 1987 (see Table 5.4). The components of NRW, from field studies and observations conducted by MWSS in 1982, include (1) leaks in the mains, pipes, gate valves, hydrants, public faucets, and water

TABLE 5.4   Estimates of Nonrevenue Water

Year	NRW (%)	Year	NRW (%)	Year	NRW (%)
1973	56	1978	46	1983	53
1974	50	1979	47	1984	54
1975	50	1980	47	1985	61
1976	50	1981	49	1986	66
1977	48	1982	52	1987	61

meters, 55 percent; (2) underregistration of customer meters and under-estimation of consumption by unmetered customers, 20 percent; (3) illegal and unauthorized use, 15 percent; and (4) public or free use, 10 percent.

Another set of data, emphasizing the discrepancy between the actual volume of water produced and distributed against that actually billed, is presented in Table 5.5. From 1981 to 1985 there was a consistent decline in the percentage of water billed. Whatever the factors contributing to this phenomenon, the record implies the urgency of curbing such losses.

Currently meter reading is undertaken by about 150 MWSS employees from nineteen branch offices. Computerized bills are issued monthly. However, collections are decentralized and are contracted to individuals or companies directly responsible to the branch offices. In 1985, collection efficiency of MWSS declined to 85 percent because of economic recession and bankruptcies. About ₱154.6 million, representing 50 percent of total collection of MWSS, is due from government agencies.

Measures are being taken to upgrade the billing and collection system, and a pilot project on improved billing methods will be implemented soon. There is also an intensified campaign to search for illegal connections, legalize unauthorized water and sewer connections, improve meter reading, and install meter seals.

TABLE 5.5   Comparison Between Water Produced and Distributed and Volume Actually Billed (m³/sec)

	1981	1982	1983	1984	1985
Water produced and distributed	493.39	596.54	625.80	677.40	797.35
Water billed	271.06	269.51	287.45	289.34	302.68
Percentage of water billed to produced	54.94	45.18	45.93	42.71	37.96

Source: MWSS 1986.

TABLE 5.6   Water Production and Consumption (million m³)

Year	Water Produced	Domestic	% of Production	Commercial	% of Production	Industrial	% of Production
			Consumption				
1984	667.80	168.55	25.24	106.40	15.93	14.95	2.24
1985	786.83	183.55	23.33	104.84	13.32	14.46	1.84
1986	904.51	195.47	21.61	100.79	11.14	14.52	1.61
1987	862.62	218.48	25.33	101.76	11.80	16.27	1.89
1988	878.82	225.85	25.70	112.71	12.83	20.90	2.38

The Water Demand Situation

Components of water consumption in Metropolitan Manila are the domestic, industrial, commercial, and institutional sectors. About 60 percent of the total water consumption from MWSS is for domestic use. Domestic consumption ranges from 170 to 780 ℓ/capita/day.

Table 5.6 indicates the comparative record for the three main water-using sectors in Metropolitan Manila from 1984 to 1988. Considering the actual water production, the data indicate a very low volume (34.36–43.41 percent) of water used against production. For the same period, the volume of nonrevenue water ranges from 59.07 to 65.64 percent.

The demand for water is expected to increase throughout the years (see Table 5.7). For 1989 the demand of 2.33 million m³/day appears more than adequately met by the present volume of water produced by MWSS, about 2.4 million m³/day. Unfortunately, however, because of distribution losses the actual water supplied averages only about 1,575 million ℓ/day, or approximately 76 percent of consumption requirements (ADB 1987:19).

Allocation of water to the different sectors depends on the users' proximity to water distribution lines. Applicants for service connections in areas with existing water mains are served more promptly than those applicants without access to water distribution lines. New water connections are installed without charge by MWSS.

In 1986, the Asian Development Bank (1987:19) reported that the MWSS existing facilities are generally in poor condition and that the majority of water mains are inadequate in size to meet actual demands. In some areas, individual service connections are 100 m or more from the nearest water main to consumers' properties. These, locally known as "spaghetti connections," are prevalent because of the low density coverage of the distribution system. Consequently, water pressure throughout

TABLE 5.7   Projected Demand for Water

Year	Estimated Population (million)	Water Demand (million m³/day)
1987	8.16	2.17
1988	8.40	2.25
1989	8.64	2.33
1990	8.89	2.42
1991	9.11	2.49
1992	9.34	2.57
1993	9.57	2.64
1994	9.81	2.71
1995	10.05	2.78
1996	10.26	2.85
1997	10.48	2.93
1998	10.70	3.00
1999	10.92	3.09
2000	11.15	3.17

Source: MWSS 1986.

the majority of the system is inadequate during the day when water use is at its peak.

Water Quality

To ensure that water supplied is safe for the health and well-being of consumers, MWSS regularly monitors water quality through sampling and analysis of water from river sources, treatment plants, points in the distribution system, artesian wells, and water samples submitted by both government and private establishments.

The MWSS claims that the water harnessed from the Angat, Ipo, and La Mesa dams is treated and transformed into one of the purest, cleanest, and safest in the world, surpassing even international health standards (MWSS 1989b:1).

The lead agency for drinking water quality surveillance and control is the Department of Health. The quality of the treated water in the distribution network is monitored not only by the MWSS, but by an inter-agency committee called the Metro Manila Drinking Water Quality Committee. This committee is composed of the Department of Health— Bureau of Research and Laboratories; National Capitol Region for Health, Department of Environment and Natural Resources (DENR)— Environmental Management Bureau, Manila Health Public Laboratory; and the MWSS. The Technical Committee meets monthly to evaluate the results of bacteriological analyses of water samples. Efforts at improving water quality are directed at preventing contamination resulting from cross-connections, back-siphonage, leaky connections, defective storage tanks and service reservoirs, low water pressure, and use of booster pumps and pressure tanks by customers.

Approaches to Improve
Water Management and Conservation

Progressive Rate Structures

Implicit in the MWSS tariff rates is the intention to manage water demands. By this assessment, MWSS hopes that excessive and wasteful water uses can be minimized, so that more water will be available for productive purposes. The tariff scheme includes various scales of fees that increase with the volume of water consumed (see Table 5.8). Domestic users have the lowest charges, and the industrial sector the highest. Equity considerations are the basic reason for this.

For a family of eight living in the central service area, and with an average consumption of 15 m^3 per month, the combined water and sewer charge amounts to ₱30.00 per month. Besides the basic water charge, the MWSS tariff consists of environmental/sewer charge, currency exchange rate adjustment, and maintenance service charge. The currency exchange rate adjustment is included to cover possible losses arising from foreign exchange fluctuations, which is to be expected because of heavy borrowings by MWSS from foreign banks. The maintenance service charge, which usually ranges from ₱1.50 to ₱50.00 per month, is levied to defray the costs of repairs and maintenance of water meters. This service charge varies according to the type of consumer.

Increasing water demand has prompted the MWSS to formulate long-term and short-term improvement plans, centered on two approaches: reducing the levels of nonrevenue water and developing new supply sources (Rew 1976:6–7).

Reduction of Nonrevenue Water (NRW)

During the second half of the 1980s, the MWSS engaged in two projects to lower NRW to an acceptable level. The first of these was the Manila Water Supply Rehabilitation Project (MWSRP I). This project aimed at halving NRW from the 1984 rate of about 54 to 25 percent by 1992. Phase I of the project, started in December 1984 with substantial support from the Asian Development Bank, involved intensive physical rehabilitation of the distribution system in fifty-six zones of the most densely populated areas, and upgrading of the maintenance function of the MWSS. Phase II, started in April 1988, covered about 35 percent of the total MWSS supply including groundwater.

TABLE 5.8 Schedule of Basic Water Charges By Consumer Category (effective October 1, 1989)

Type of Service/ Consumption Range (m³)	Water Rate/m³ (₱)
Residential A	16.00/connection
First 10	2.15
Next 10	2.70
Next 10	3.25
Next 10	3.80
Next 10	4.35
Next 20	4.90
Next 20	5.45
Excess over 100	6.00
Residential B	
First 10	17.00/connection
Next 10	2.25
Next 10	2.80
Next 10	3.35
Next 10	3.90
Next 10	4.45
Next 20	5.00
Next 20	5.55
Excess over 100	6.10
Commercial	
First 25	152.50/connection
Next 975	6.10
Excess over 1000	6.40
Industrial	
First 25	170.00/connection
Next 975	6.80
Excess over 1000	8.15

Source: MWSS 1989a.

The second project was the Metro Manila Water Distribution Project, which aimed to distribute the improved water supply from existing and recently completed sources to areas that can be economically reached by the central distribution system, including those not previously covered in the design of MWSRP II. Through this project, the MWSS hoped to generate revenues to help meet its increasing debt-servicing requirements. The project was to involve installation of 100,000 house service connections and 207 public standpipes, primarily in subdivisions, pockets of infilling developments, and blighted areas.

Development of New Supply Sources

In order to maintain a policy of "no water shortages" despite the huge financial investment requirements, the MWSS has also continued to develop new water supply sources. For example, the Manila Water Supply Project II, built from 1977 to 1986 at a total cost of ₱4.5 billion, sought to provide additional potable water from improved utilization of the existing Angat-Ipo-La Mesa-Alat sources, developing additional groundwater sources, and rehabilitating the distribution system. The Angat Water Supply Optimization Project sought to provide water to an additional 360,000 service connections through (1) improvement of the Angat Auxiliary Hydroelectric Plant; (2) construction of one intake structure at Ipo, one 6.4 km long tunnel from Ipo to Bicti, and one 16.3 km long aqueduct from the Bicti to the La Mesa Reservoir; (3) expansion of La Mesa Treatment Plant; and (4) construction of an additional distribution network of various pipe sizes, connections, and boosters in the MWSS area. The MWSP III, a 10-year project expected to be completed by 1997, is designed to tap the Kaliwa Basin as a new source for water supply and incidental power, through the construction of a large dam and diversion tunnels, yielding an estimated increment of 1.9 million m³/day.

In addition to the preceding water supply projects, other potential sources are being considered for future undertakings. For instance, the Kanan River Basin is planned for integration under a transbasin arrangement with MWSP III. The Kanan River has a potential yield of about 3.2 million m³/day. Another source, the Umiray River Basin with a potential yield of about 777,000 m³/day, is likewise being considered in future expansion work. Other sources include Laguna de Bay, the Marikina River, the Pampanga River, and Taal Lake in Batangas with a total potential yield of approximately 6.3 million m³/day. Even without these water resources, the MWSS foresees that the water supply will still be more than adequate in the years subsequent to 1990.

At present, conflicts from the allocation of water for power generation, irrigation, and urban uses are hardly apparent. However, the tremendous growth in population, urbanization, and improved agricultural technology might, in the future, require more deliberate and stringent allocation mechanisms.

Which sector will be given priority is an issue that must be resolved by higher authorities. MWSS officials have revealed that at least for now, their agency gets priority over the allocation of water even during dry periods. Furthermore, out of the total yield of 56 m³/sec from Angat Dam, 22 m³/sec is allocated for the MWSS, and 34 m³/sec is for power generation. Since increasing energy shortages may influence the volume of water for domestic purposes, water allocation mechanisms may have to be reviewed.

In the future, balancing of the sectoral demands will become a crucial issue. Resolution by analysis and deliberations should not be too difficult, because of the coordinative authority of the National Water Resources Board as mandated in the Philippine Water Code.

Watershed Protection and Conservation

Sustainability of surface water supply for Metropolitan Manila is very much a factor of the integrity of the watershed areas. Safeguarding the catchment areas is a critical component of overall water resource management.

The most important watershed is that of the Angat River, which supplies water for the multipurpose Angat Dam. Recently, illegal logging activities in the watershed were the subject of Senate Committee hearings. The destruction of the watershed area of Angat has been blamed for the frequent and prolonged brownouts in Metropolitan Manila. Business owners and industrial operators expressed concern over the economic implications of power disruptions (*Manila Bulletin* 1989:1, 20).

Recognizing the impacts of the population's activities in the area, the government introduced development programs in the catchment. Of special interest is the Integrated Social Forestry Program of the Department of Environment and Natural Resources, which seeks to provide livelihood opportunities to upland farmers while conserving the ecological stability of the natural catchment. However, there is an urgent need to upgrade the administrative planning and implementation capability of the program (Fellizar 1987:320).

Other projects are being implemented in the watershed, including reforestation and streambank stabilization. Despite these projects, the more pressing concern is to regulate migration into the basin. Failure to do so in the short term will create detrimental consequences to the concern of providing a sustainable water supply to Metropolitan Manila.

The Angat watershed is under the administrative jurisdiction of the National Power Corporation, and the Umiray River is under the management of the Department of Environment and Natural Resources. Basin areas for the Alat and La Mesa rivers are under the responsibility of the MWSS.

Recommendations

Technical, financial, organizational, and sociopolitical issues must be addressed by the MWSS and the government. For this undertaking, the following steps are recommended:

1. **Establish a desirable balance between efficient water distribution and supply development.** Decisions must be made either to invest more loan money on supply development projects or to improve the existing distribution system. The issue is not simply one of increasing supply but equally that of making efficient use of the current supply level. Losses from leakages, measurement errors, and improper connections are technical concerns that need fast and drastic measures. Optimizing the use of available supply is a desirable option, socially and economically.

2. **Enhance the financial viability and organizational efficiency of MWSS.** Maintaining financial viability of the MWSS is a challenge. This concern is of paramount value considering the amount of monetary investment required in its development efforts and the equally pressing need of servicing its foreign loans. Maintaining an acceptable level of financial soundness requires improvement in the organizational efficiency of the MWSS. Specific measures will have to be instituted to reduce illegal connections, incomplete records, billing errors, and proliferation of unregistered meters. Also, there is a need to increase the billing and collection rates. Training of personnel, adequate compensation, and alternative institutional arrangements are some of the interventions to be considered. A sound financial scheme for debt-servicing is in order, one that will not penalize the poorer sector of society.

In addition, increasing the revenues of MWSS would require servicing a greater number of still-unreached sectors. Although more capital investment will be required, it will result in greater satisfaction among the consumers.

3. **Encourage consumer participation.** One dimension not given much attention in water resource management for metropolitan areas is public participation. Consultation, dialogues, and voluntary services are hardly resorted to. The role of consumers in shaping water supply policies and directions has yet to be explored. This has great potential; it only has to be given a chance to work. Consumers' involvement in the formulation of possible alternatives to the present water distribution system and other relevant plans is essential to ensure consumer support and satisfaction. Making the public more aware of the plans, priorities, and problems associated with water management will minimize complaints and may create a desire for more meaningful participation in water-related projects. Dissemination of information to heighten water conservation consciousness among people is a must. The people themselves can perhaps institute self-policing mechanisms whereby illegal connections, pilferage, and other wasteful practices can be eliminated. For instance, public support is needed to minimize, if not totally eliminate, the unregulated use of fire hydrants.

4. Harness the local governments for water management concerns. Local governments are formal structures whereby citizen participation in resolving issues related to water management can be channeled. Surprisingly, local government units in the metropolis have not been actively involved in the issues of water management. Provision of water through a public corporation such as the MWSS has somewhat diminished the initiatives of the local governments and the consumers to engage in matters concerning water use. Provision of adequate and safe water is not an exclusive responsibility of the MWSS. The local governments within the metropolitan area have powers and resources that can be tapped for the planning and implementation of water-related programs within their jurisdictions. Local governments must be given their share of responsibilities and benefits as rightful partners in water management. This implies a clear role perception, as well as commitment to a common goal, and joint efforts.

5. Prescribe incentive system for efficient water use. A mechanism has to be institutionalized whereby judicious use of water in the industrial, commercial, public, and residential sectors is recognized and appropriately rewarded. Such an incentive will somehow develop water conservation consciousness among consumers, in addition to realizing more profit for MWSS. In this manner, more water will be made available to consumers for greater revenues.

6. Integrate water management concerns in overall urban development. Water availability is at the core of urban welfare and productivity. Therefore, concern for water should be an integral component of urban development planning to avoid unforeseen problems from occurring. Policymakers and planners tend to isolate water issues from the total development agenda. A policy, which must be formulated to effect such changes in development perspectives, calls for support from the national and local leadership.

7. Upgrade the management information system. Accurate and reliable information is unquestionably of paramount value in planning and decision-making activities. Metropolitan Manila severely lacks such information, which is essential for providing a basis for sound demand-and-supply forecasts. Even where data can be found, there are enormous problems (Palencia 1984:192–196). For instance, statistics are compiled for different purposes; identifying the agency that handles the desired information is a difficult task; and discovering that the same information was processed using varied estimates and assumptions is discouraging. To correct these problems and to provide accurate information to planners, decision-makers, and the general public, the system needs a framework by which relevant data can be systematically collected, organized, and analyzed.

References

ADB (Asian Development Bank). 1987. Philippine Water and Sanitation Sector Profile. Manila, Philippines.

Fellizar, Francisco P., Jr. 1987. Integrated Social Forestry as a Development Program: An Assessment of Administrative Capability, Integration, and Effectiveness. Ph.D. dissertation, University of the Philippines, Manila.

IEP (Institute of Environmental Planning). 1971. Manila Bay Metropolitan Framework Plan. University of the Philippines, Quezon City.

Manila Bulletin. November 22, 1989. Senate Probes Angat Logging. Manila, Philippines.

MWSS (Metropolitan Waterworks and Sewerage System). 1986. Annual Report. Manila, Philippines.

———. 1989a. The Metropolitan Waterworks and Sewerage System Through the Years. Public Information Department, Philippines.

———. 1989b. MWSS Plans and Programs. Manila, Philippines.

NEDA (National Economic Development Authority). 1982. The 1982 Philippine Statistical Yearbook. NEDA Publications Office, Manila.

NWRC (National Water Resources Council). 1976. Philippine Water Resources. Manila, Philippines.

———. 1980. Groundwater of the Philippines. Manila, Philippines.

———. 1982. Philippine Water Code and Implementing Rules and Regulations. Manila, Philippines.

Palencia, Lamberto C. 1984. Water Supply Planning for Metro Manila: Some Economic Considerations. Ph.D. dissertation, University of Hawaii, Honolulu.

PICOREM. 1980. Metro Manila Water Supply III: A Report. Manila, Philippines.

Rew, Alan. 1976. "Access to Urban Water: Social and Institutional Dimensions of an Environmental Issue." *Philippine Planning Journal* 8(1): 6–7.

6

Water Resource Management in a Metropolitan Region Downstream of a Large Lake: Osaka, Japan

Michio Akiyama and Masahisa Nakamura

One of the main challenges to water resource management in postwar Japan has been the rapid increase in urban domestic and industrial water demands. To meet these demands, the urban sector relied on the construction of multipurpose dams, particularly in the 1950s and 1960s. This "structural" approach allowed new water users to avoid direct confrontation with the traditional water users in the agricultural sector. By tapping previously inaccessible sources of water, they created new water rights for domestic and industrial uses. Yet to avoid confrontation over the well-established rights of irrigators to virtually all of the minimum flows in the rivers, urban water users had to expand the geographical scope of their water management systems more than they would have had to otherwise.

The Osaka Metropolitan Region is typical in Japan in that it has had to cope with rapid urbanization and industrialization. Yet it is also unique in that the Yodo River, the only natural outlet from Lake Biwa, the largest water source in Japan, flows through the center of the region (see Figure 6.1).

This paper presents an overview of the evolution of water management issues in the Greater Metropolitan Osaka Region, focusing on the Yodo River–Lake Biwa system. It begins with some background information on Osaka, the second largest megalopolis in Japan, and on the Yodo River, which has sustained its urban and industrial activities. The main section of the presentation consists of two parts: (1) an historical overview of water resource development in the lower reaches of the Yodo River basin and its relationship to the entire system, and (2) an

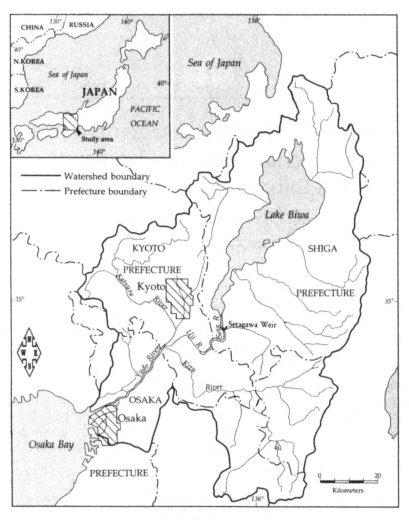

FIGURE 6.1 Lake Biwa and the Yodo River region (*Source:* Shiga Prefectural Government, 1985:3).

illustration of the dynamics of management in the metropolitan region with particular reference to the development of urban water supply systems.

Osaka and the Yodo

Although many Japanese associate the Yodo River with Osaka, the area officially designated as the Yodo River basin includes several upstream geographical and administrative regions. The headwaters, almost all in Shiga Prefecture, consist of Lake Biwa and the rivers flowing into it. The single river flowing out of the lake is called the Seta until it reaches the boundary of Kyoto Prefecture. Then it becomes the Uji River until it joins the Kizu and Katsura rivers near the boundary of Kyoto and Osaka prefectures, 50 km downstream of the southern edge of Lake Biwa. The result of this confluence, the Yodo River, flows through Osaka Prefecture and Municipality into Osaka Bay.

The entire catchment area is 7,281 km² above Hirakata, the principal measuring point reference about 20 km upstream of the river mouth. The annual average flow, high flow, and low flow of the Yodo are, respectively, 177.6, 226.8, and 117.0 m³/sec. The ratio of high-to-low flows, 1.94, is the lowest among the major river systems in Japan.

One reason for this stability is the difference in meteorological and runoff characteristics in the three upstream river systems—low flows and high flows rarely coincide. In addition, the numerous lowland areas in the upper basin of Yodo River function as natural underground reservoirs that provide stable water replenishment of the lower stretch of Yodo River. The outflow of the Yamashiro (Kyoto) Lowland Area is almost 10 m³/sec during droughts. The Greater Metropolitan Osaka Region, which consists of Osaka City and the forty-three municipalities surrounding it, coincides roughly with Osaka Prefecture. It is the largest urban and industrial center in western Japan. The population of Osaka City is approximately 2.6 million. In the Meiji Era, from 1868 to 1912, Osaka thrived as the commercial center of Japan, and industrial activity exceeded that of Tokyo in the prewar years. After the war, however, the major commercial operations in the Osaka region moved their headquarters to Tokyo for better access to critical information for strategic corporate decision-making. The trend has never reversed, leaving Osaka unable to rebuild its former core of industries—heavy manufacturing, allied chemicals, and textiles. The gap has widened because of continuing centralization of international financial and information activities in Tokyo. Nonetheless, Osaka remains the leading urban industrial center west of Tokyo, and there have been some successful attempts in recent years to regain its economic vitality. Current projects to construct a new off-shore international airport in Osaka Bay and a science-park city in Nara are examples of promising attempts to restore Osaka (Edgington 1990).

Water Resource Development in the Yodo River System

In Japan, management of sixty-four principal river systems was brought unilaterally under government control in 1937. The first stage of the water control project on the Yodo River was initiated in 1943 and completed in 1952. Under this scheme, management of the river was to be improved first by changing the operating methods of the Setagawa Weir to lower the lake level by 1.3 m. The incremental Yodo River water rights, totaling 136.62 m^3/sec, were allocated, based on existing water use patterns, to agriculture (16.8 m^3/sec), drinking water supply (23.2 m^3/sec), industrial water supply (8.12 m^3/sec), and low-flow augmentation (88.5 m^3/sec). However, questions were raised as to the unintended impacts of the lowered water level on various existing structures and facilities at Lake Biwa, as well as to the fishery activities in the lake, at that time a major industry in Shiga Prefecture.

The nearly two-thirds of the supplemental allocation reserved by the Ministry of Construction for low-flow augmentation has also been controversial. Originally the ministry contended that the 88.5 m^3/sec was necessary for navigation, but shifted its justification in the 1970s to maintenance flow to preserve water quality. The ministry has refused to allow the low-flow augmentation to be reduced for transfer to other purposes, despite occasional calls for it to do so.

Industrial water demands began to increase sharply as the country entered an era of economic growth a decade or so after the end of World War II. In the downstream stretch of Yodo River, the thriving Hanshin Industrial Belt established in prewar years thirsted for more water. Exploitation of groundwater, the most common means for meeting the growing industrial water demands then, soon became constrained due to competition among industrial establishments and to land subsidence caused by overdraft of water, particularly in the lowland of Osaka and in Amagasaki, the coastal region immediately west of Osaka. Industries were then forced to look for alternative sources of water. The concept of an industrial water supply system, unique to Japan in terms of scale, was developed in 1958. The first of a series of such systems in the country was introduced in the Amagasaki-Nishinomiya region.

The industrial water supply systems allowed Japan's rapid industrialization in the 1960s to occur without necessitating a change in the existing water rights structure. It would have been largely impossible to provide water rights from existing river systems to the large complexes of water-intensive industries such as steel, pulp, cement, and chemicals that led the growth surge. The industrial water supply systems, established for the industrial complexes and administered by the Ministry of International Trade and Industry, developed new water rights through constructing dams to capture wet season "surplus" waters.

The first modern municipal water supply system within the Yodo River system was established in Osaka in 1895, with a design service population of 610,000. Six expansion projects were constructed in the period up to World War II. The city of Kyoto initiated a water supply service in 1912, taking advantage of its access to Lake Biwa water carried through a famous tunnel diversion, Biwako Sosui. In 1926, 1927, 1934, and 1940, the suburban cities of Neyagawa, Amagasaki, Moriguchi, and Hirakata, respectively, began to depend on the Yodo River for their water supply sources. Furthermore, two public water supply utilities—Hanshin Water Supply Utility and Osaka Prefectural Water Supply Utility—obtained permits for access to the Yodo River for drinking water. They planned to wholesale the treated water to small municipalities in suburban Osaka.

The demand for municipal water supply began to increase significantly in the 1960s throughout Japan due to population increase, expansion of service coverage, and increases in per capita consumption. This trend was even more striking in large metropolitan areas like Osaka where water-consuming municipal activities are much more varied and intense than in smaller cities.

As mentioned earlier, the industrialization of Japan resulted in serious conflicts of water use with the traditional agricultural sector throughout the 1950s and 1960s. These conflicts were solved by the introduction of additional sources of water through newly constructed multipurpose dams. The situation was different in the Yodo River basin, where the share of agricultural water demand is relatively small. Instead, the conflicts were between the upstream and downstream interests.

For centuries the communities immediately surrounding Lake Biwa experienced many severe floodings of their agricultural fields. The government was reluctant to undertake major dredging work on the Seta River at the outlet of the lake primarily because of opposition from downstream communities who feared flooding of their paddy fields. Finally, about a century ago, the government undertook a major dredging along with construction of the Setagawa Weir.

By the mid-1950s, potential downstream beneficiaries became interested in a comprehensive development plan for Lake Biwa water resources. Although various downstream governments and industrial interests were demanding more water, the upstream water users in Shiga Prefecture were reluctant to let the lake water level be lowered to accommodate those demands. The Lake Biwa Comprehensive Development Committee was organized in 1956 with representatives from the Ministry of Construction, Shiga Prefectural Government, Osaka Prefectural Government, Hyogo Prefectural Government, City of Kyoto, City of Osaka, Hanshin Water Supply Utility, and Kansai Electric Company. The committee proposed construction of a barrier separating the northern and southern lakes to allow the southern lake to be drawn down while maintaining the water level in the northern lake. Several alternatives involving major construction in and around the lake were subsequently suggested by various other agencies. Finally, in the late 1960s, the prototype of the current Lake Biwa Comprehensive Plan, keyed to improve the discharges and to control capacity of the Setagawa Weir, was developed. The Special Act for Lake Biwa Comprehensive Development was enacted in 1972 to implement this plan.

Although construction of multipurpose dams was about the only means in other parts of Japan to increase water availability to accommodate industrial and other municipal water demands in the postwar decades, Osaka and the surrounding municipalities benefited from their location downstream of Lake Biwa. Control of the lake water level by the Setagawa Weir had about the same effect for them as construction of a major reservoir upstream. Although the Yodo River Water Control Project in 1943 proposed lowering the water level by 1.0 m, the Comprehensive Development Plan called for lowering of water level by up to 1.5 m to develop additional water rights totaling 40 m^3/sec by 1990. Distribution of these rights would be 27.2 m^3/sec to domestic water supply (22.5 m^3/sec to Osaka and 4.7 m^3/sec to Hyogo prefectures) and 12.8 m^3/sec to industrial water supply. In exchange, major compensatory public works projects were to be built for the communities in the immediate vicinity of the lake, together with projects to counter adverse impacts caused by the lowered water level.

The Postwar Relocation of Industries in the Osaka Region

The areal extent of a metropolitan region depends on the type of metropolitan functions and the distance from the center of the metropolis. The area often extends over many jurisdictional boundaries for many functions. As for Japan's water supply systems, however, it is appropriate to confine the areal extent to within a prefectural boundary, since the systems rarely cross prefectural boundaries.

As of 1989, Osaka Prefecture had a population of 8.76 million, of which 2.63 million resided in Osaka City. The metropolitan area expanded during the 1960s and 1970s. Although recently the rate of expansion has slowed down considerably, the trend persists.

Different types of industries are located in different parts of the region. Government offices and corporate headquarters are located in the core area, together with major business establishments known to have national-scale operations, commercial establishments, and supporting service operations for business activities. Both the inner and outer rings mainly house manufacturing industries. With few exceptions, the tertiary industries in these rings primarily serve the needs of the residing population. In both the inner and outer rings, the average size of manufacturing industries has been shrinking, whereas tertiary industries have been expanding. The trend is similar in the more peripheral metropolitan area.

Two observations may be made regarding industrial relocation in the Osaka Metropolitan Region, which have significant implications in water use and management. First, relocation takes place foremost among the most rapidly growing industries. The machinery industries, which are mobile and have minimal water and land requirements, possess greater mobility. Upon deterioration of the conditions for industrial growth, they seek better opportunities elsewhere; thus, they are called "foot-loose" industries. On the other hand, the material processing industries, with bulky, usually polluting processing equipment, are difficult to relocate. Their relative immobility is because of stringent conditions for a relocation site and close association with the existing infrastructure on the current site. Thus, although water demand in the metropolitan core area has shifted from secondary to tertiary industries, many secondary water-intensive industries remain. In the metropolitan fringe area, on the other hand, the growth of population and the relocation of less water-intensive tertiary industries have become the main contributors to increases in water demand.

Second, relocation of industries, either secondary or tertiary, is accompanied in many cases by the separation of certain operations. For example, manufacturing industries leave their headquarters in the central core area and relocate production to the outer ring or the metropolitan periphery. Wholesale dealers leave the transactions operations in the central core while the distribution warehouses and transportation facilities are relocated to the outer ring. Such separations may take place in aggregation, as in industrial estates, warehouse parks, and transport terminals for manufactured goods.

Once the areal extent of a metropolitan region reaches a certain level, however, further expansion has only adverse impacts of sprawl and fragmentation.

Municipal Water Supply Systems in the Osaka Region

The water supply service coverage within Osaka Prefecture reached 99.7 percent in 1983. Water supplied to the region increased by about 1.5 times in the 20-year period since the mid-1960s, peaking in 1978. Although there have been minor changes in the 1970s and 1980s, the total water supply seems to have reached a plateau.

Water supply for Osaka City steadily increased until 1972 and has been decreasing rather steadily since. However, the aggregate water demand of other municipalities in the prefecture has been steadily increasing, except perhaps in 1979 and 1980. The city demand, which comprised more than two-thirds of the total in 1965, was overtaken by the suburban municipal supplies in 1974. The two trends cancelled each other out to result in the plateau for the whole region mentioned earlier.

Increasing demand was met basically by expanding treatment and distribution capacities rather than by introducing measures to curtail it. The expansion continued until the early 1970s, but not since, as the demand fell.

The expansion of the suburban water supplies to surpass the capacity of the city means that the regional water supply profile, once dominated by the Osaka City water supply, is now controlled by the suburban water supplies. Expanding the water supply capacity requires access to water sources as well as construction of facilities for water distribution. Although individual municipalities are responsible for providing water, not all of them are capable of providing for rapidly growing demand, particularly for gaining access to new water sources. The expansion of urban areas and the resulting increase in water demand, therefore, invariably led to expansion of a regional water supply system.

In the prewar years Osaka City provided water out of its boundaries to meet the growing demand in suburban municipalities unprepared to accommodate their water needs. Thereafter, the Osaka Prefectural Government assumed the responsibility for facilitating water to unprepared or marginally prepared suburban municipalities. Thus, in essence, the region-wide water supply undertakings were promoted by the Osaka Prefectural Government.

Although rarely done, prefectural governments are authorized to provide water supply services in Japan. Although the original Water Supply Ordinance enacted in 1890 stipulated that the responsibility to provide water supply services lies within municipal governments, the provision to allow participation of other entities into water supply services was introduced in the second revision of the Ordinance in 1913. The assumed participation under this provision, however, pertained to private entities rather than to prefectural governments. The chiefs of the Sanitation and Civil Engineering bureaus of the Ministry of Interior responded to the inquiry by the Kanagawa Prefectural Governor in 1930 that prefectural governments may be authorized to undertake water supply services only when the relevant municipalities formed a region and the total construction cost of individual systems was more than the cost of a single system managed by the prefecture.

Osaka Prefecture began construction of its water supply system in 1940. The construction was halted in 1944 because of World War II and restarted after the war. The system was completed in 1951. The original beneficiaries were Moriguchi City, Kadoma City, Higashi-Osaka City, Yao City, and Sakai City. These cities received service from Osaka City prior to 1951. In effect, the prefectural government assumed responsibility from Osaka City as the caretaker of water supply to these and other cities in the region. The capacity of the prefectural water supply increased tenfold in the 20 years since 1960.

Water Supply Systems and National Regionalization Policy

In the postwar period, a number of cases arose in which a water supply system provided service to more than one municipality. This is called regionalization of water supply systems.

In general, the regional water supply systems consist of two components—retail (delivery to customers) and wholesale (production of treated water for retail service units). There were seventy-seven service systems in 1988, of which twenty-four were operated by prefectural governments. Only four systems were established prior to 1955, including Osaka Prefecture. Most were established after 1966, in places where water demand grew during rapid industrialization.

The major reasons given for regionalization include imbalance in demand and supply, differences in service among neighboring systems, and instability in supply capability. Financial and fee collection issues, on the other hand, were not regarded as major reasons. Access to stable water sources was by far the most important impetus toward regionalization.

The Ministry of Health and Welfare, the responsible body for national water supply policy, began to take up the issue of regionalization in the late 1960s. The Commission on Public Nuisances (*Kogai Taisaku Shingikai*) (later called the Commission on the Living Environment [*Seikatsu Kankyo Shingikai*]) issued two recommendations—"Water Supply Regionalization and Management Principles" (1966) and "The Future of Water Supply Systems and the Recommended Approach" (1973)—outlining guidelines for regionalization. The revision of the Water Supply Act in 1977 brought regionalization within a legal framework.

The Ministry of Health and Welfare considered that the lingering problems facing water supply systems two decades ago could be resolved by regionalization. The ministry categorizes the benefits associated with regionalization into "economic" and "noneconomic." The economic benefits include the lowered cost of delivered water and uniform rates over the region, while the noneconomic benefits include balancing demand and supply, controlling water pollution, enhancing safety, strengthening management and technical capabilities, and expanding services to unserved populations.

The ministry proposed a system for indicating the type of regionalization by assigning one of the following five levels of jurisdictional authorities to each of the three categories of water supply service functions (i.e., water source management, wholesale services, and end-user services): R: regional (several prefectures) level; P: prefectural level; A: area level; D: district level; and M: municipal level. For example, an RPM-type regional water supply would involve several prefectures in obtaining access to water sources, a single prefecture in the wholesale service, and municipalities in the end-user services. According to the preceding categorization, Osaka City water supply systems are either MMM or RMM types, whereas the municipal systems are RPM, DDM, or MMM types. The current thought of the ministry is to raise the lower level services under each of the three categories of service functions to a higher level.

In the Osaka Metropolitan Region, the Osaka prefectural water supply system has been regionalized. Metropolitan water supply systems in Japan faced severe shortages of water prior to the late 1960s. Many of them, exemplified by Osaka Prefecture, had already undergone regionalization by the time the regionalization scheme was proposed by the ministry.

Emerging Water-Use Conflicts

Because the Osaka prefectural system was the last to compete for acquisition of water rights, it has been most vulnerable to water management policy decisions involving water rights. As a result, there was a deficiency of water equivalent to 2.3 m^3/sec of water rights, or approximately 198,000 m^3/day, at the time of the sixth expansion, with its design capacity of 2,130,000 m^3/day.

With the Yodo River as its sole and stable source of water, the Osaka prefectural water supply system was able to expand its service area to meet the needs of suburban municipalities having difficulty in finding suitable water sources nearby. The growth of the prefectural water supply system and of the suburban municipalities were, therefore, inseparably related. At the completion of the seventh expansion project, the prefectural system will have most of the municipalities included within its service area. In the process, the prefectural system has created uniformity in service quality among municipalities regardless of their distance from the water source. Therefore, even when there were relocations of water-intensive metropolitan functions and activities out of the city districts into the suburban municipalities, it has been difficult to open up additional water sources, unlike the case in the 1960s. It is much more difficult to find suitable new water sources and dam sites, and the cost of constructing supply facilities has increased. For example, a certain degree of demand control through the introduction of a suitable pricing policy would be easier to achieve today than in the 1970s. In the past decades, much experience has been accumulated in managing the region's limited water resource.

The integration of the municipalities into the Osaka prefectural water system has one important implication—domestic demand within the region no longer needs to compete with other types of water uses, particularly agriculture. In addition, the reduction in groundwater draft has resulted in fewer conflicts over aquifer usage. The establishment of the prefectural system was in fact catalytic in overcoming water-use conflicts among municipalities.

The success in regionalization also meant the expansion of the service coverage area, in some cases beyond the watershed of the Yodo River basin. This expansion invites additional development of the river, which may aggravate upstream-downstream conflicts. In other words, resolving conflicts among different users within a given metropolis may result in greater conflict with other communities and regions in the river system.

The issue of economic efficiency versus equity in water supply in the Osaka Metropolitan Region remains unresolved.

The regionalized water supply, particularly with expanding service coverage of the Osaka prefectural water supply system, contributed to resolving the current differences in water charges among municipalities within the region. But at the same time, demands for water suppressed by limited supply capacity and high charges, particularly in municipalities depending on their own water supply systems, would increase with sufficiency in supply capacity and lowered charges brought about by regionalization. On the other hand, upstream-downstream conflicts continue to highlight the importance of demand management (i.e., a reduction in waste and promotion of reuse and recycling, which are associated with locally appropriate water resource management). Osaka Metropolitan Region now faces this issue as the Lake Biwa Comprehensive Development Plan, giving Osaka the right to draw more water from Lake Biwa, is to be put into practice.

Conclusion

The postwar Japanese water resources development activities took place basically in three distinctly different stages (Takahashi 1988). The first, corresponding to the period between 1945 and 1959, focused on prevention and management of catastrophes like floods. The second, between 1960 and 1972, coincides with the period of unprecedented economic development of Japan. At that time, water users faced severe water shortages, and many major water resource development projects were carried out. Because of the long project duration involving civil engineering construction works and development of institutions for supporting them, the impact of policy decisions made during this period has lingered. The third stage, 1973 to date, corresponds to sophistication of water resource management, with the slowing down of the global as well as the Japanese economies. Regionalization has reduced uncertainties and inequities in supply. Partly due to the policies of levying progressive charges on metropolitan water users and wastewater dischargers, many industrial and commercial establishments have successfully adopted water reuse and recycling technologies.

The Yodo River region, with its unique water source in Lake Biwa, has also reflected national shifts in economics and in water management policy.

References

Edgington, D. W. 1990. "Managing Industrial Restructuring in the Kansai Region of Japan." *Geoforum* 21:1–22.

Shiga Prefectural Government. 1985. *Lake Biwa: Conservation of Aquatic Environments.* Environment Division, Otsu, Japan.

Takahashi, Y. 1988. "Toshi to mizu" (Cities and water) (in Japanese). *Iwanami Shinsho*, No. 34, Japan.

7

Water Use Conflicts in the Seoul Metropolitan Region

Euisoon Shin

Seoul has been the capital of Korea for the past 600 years. The population of Seoul reached 1 million in the 1940s and was 2.4 million in 1960. Since then, it has grown more than twice as fast as that of South Korea as a whole, partly because of rapid industrialization. By 1989, the population of Seoul exceeded 10 million, or 24 percent of the total population of South Korea.

The Han River, one of four large rivers in South Korea, flows through Seoul Metropolitan Region. Until the early 1970s, the Han River provided sufficient clean water for farms and households in the river basin. However, rapid industrialization and urbanization in the region increased water use in the domestic, commercial, and industrial sectors, causing major problems.

First, as the increasing demand could not be satisfied with the natural flow of the Han River, multipurpose dams were constructed to increase the potable water supply. These facilities are expected to meet water demand in the Seoul Metropolitan Region until the year 2000. Most multipurpose dam sites have already been developed, however, and even if groundwater can help alleviate upcoming water shortages, its effect will be minimal. That means water resource management in the region should include demand restriction as well as supply expansion in the future.

Second, people and industries along the Han River and its tributaries have discharged sewage and wastewaters beyond the river's self-purification capacity, polluting the river. As a result, Seoul was obliged to purchase more raw water from Paldang Reservoir, which is located upstream on the Lower Han River. The water quality of Paldang Reservoir has also been deteriorating recently. The reservoir may become unsuitable as a source of city water unless proper measures are taken soon.

This chapter first looks at the water demand and supply management in the region, focusing on how the increasing water shortage problem has been resolved. It then investigates the present status of Han River pollution and efforts taken by the government to alleviate the causes and effects of water pollution. Thereupon it analyzes the water-related government organizations and institutional arrangements established for managing water resources in Korea. A summary concludes the analysis.

Water Demand and Supply Management

Water Demand and Supply in Korea

Average annual precipitation in Korea is 1,159 mm, which is 1.5 times the world average of 730 mm. However, annual per capita volume of water from precipitation is only 10 percent of the world average because of high population density. The total volume of water available in 1987 was about 114 billion m^3. Of this, 42 percent was lost through seepage and evaporation, and the remaining 66.2 billion m^3 flowed into the rivers. However, 40.5 billion m^3 was concentrated during the flood season and only 25.7 billion m^3 constituted normal flow. The surface water actually used was 16.7 billion m^3, only 15 percent of total volume. In addition to use of the surface flow, 6.5 billion m^3 was recovered through dams and 1.6 billion m^3 was recovered from groundwater. As a whole, 24.8 billion m^3, or 22 percent of total water volume from precipitation, was actually used in 1987 (Ministry of Construction 1988).

Table 7.1 shows water demand and supply trends in Korea. Water demand includes domestic, industrial, and agricultural sectors and the volume required to maintain river flow throughout the year. Total water demand has increased 26.7 percent from 1981 to 1986 and is expected to increase at a slightly lower rate in the future. In 1981, the agricultural sector accounted for more than one-half of total water consumption, and the domestic and industrial sectors used 21 percent. During the past 10 years, the agricultural share decreased relative to the domestic and industrial sectors. Water demand in the domestic sector increased 70 percent from 1981 to 1986. The industrial sector is expected to experience the highest demand growth in the 1990s.

In 1981, 19 billion m^3 of water was supplied, but 1,348 million m^3 more was required to satisfy total demand. River flow provided 75 percent of total water supply; reservoirs, 18.4 percent; and groundwater, only 6.1 percent. As stated earlier, normal river flow carries only 22 percent of the total water resource; thus, there is a limit to dependence on natural flow

TABLE 7.1   Water Demand and Supply in Korea (million m³/year)

Sector	1981	1986	1991	2001
Total demand	20,327	25,760	31,770	37,710
Domestic	2,503	4,260	5,669	7,443
Industrial	1,778	2,127	2,906	4,130
Agricultural	10,684	11,804	13,482	15,794
Maintenance	5,362	7,569	9,713	10,343
Total supply	18,979	23,685	30,211	38,503
River flow	14,319	16,359	18,265	21,404
Groundwater	1,164	1,513	1,744	2,025
Reservoir	3,496	5,813	10,202	15,074
Final balance	−1,348	−2,075	−1,559	793

Source: Ministry of Construction 1987b.

as a source of water. It is projected that increasing water demand will mostly be met by constructing new multipurpose dams. Reservoirs will provide nearly 40 percent of total supply by 2001, and the contribution of river flow will decrease from 75 to 56 percent. Groundwater will remain a minor source, as there are no large aquifers in South Korea.

Water Use in the Han River Basin

The Han River system is composed of the North Han, the South Han, and the Lower Han rivers (see Figure 7.1). It is 470 km long and the basin area is 26,219 km², covering one-third of South Korea. Total volume of water carried annually in the Han River is 32.2 billion m³, or 28.2 percent of the national total. The North and the South Han rivers originate from the mountainous area in the eastern part of the Korean peninsula and converge at Yangsuri, forming the Lower Han River. The Lower Han River flows through the fertile Kyunggi Plain and Seoul Metropolitan Region, then into the Yellow Sea.

Until the early 1970s, the Han River could satisfy water demand in the river basin with natural river flow, even during the dry season. Irrigation was the major use. However, rapid industrialization and concentration of population in the Seoul Metropolitan Region quickly increased water demand, and natural flow was no longer sufficient. To meet the increasing water demand in the Seoul Metropolitan Region, the Soyanggang Multipurpose Dam was constructed on the North Han River (1967–1973), and the Choongju Multipurpose Dam on the South Han River (1980–1985)— with a capacity of 1,213 and 3,380 million m³, respectively. In addition to these, there are six dams on the Han River, mainly for electric power generation. Paldang Dam, which is located where the North and the South

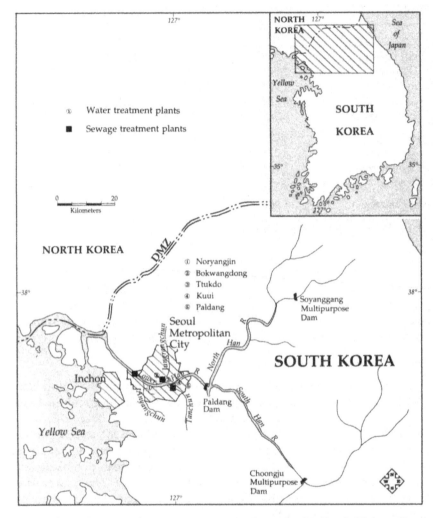

FIGURE 7.1 Water and sewage treatment plants in the Seoul Metropolitan Region.

Han rivers converge, functions as a reservoir of raw water for the Metropolitan Region Areawide Waterworks.

Table 7.2 shows the water demand and supply trends and projections for the Han River basin. There are several differences between the water demand patterns of the Han River and the nation as a whole. First, the share of maintenance water in the Han River is nearly one-half of total demand, while it is less than 30 percent in other rivers. This is because the Han River has large reservoirs, formed after eight dams were con-

TABLE 7.2   Water Demand and Supply in the Han River Basin (million m³/year)

Sector	1981	1986	1991	2001
Total demand	5,091	7,745	8,724	9,916
Domestic	1,304	2,035	2,511	2,894
Industrial	468	492	517	570
Agricultural	1,111	1,276	1,439	1,722
Maintenance	2,208	3,942	4,257	4,730
Total supply	5,248	9,432	9,696	11,357
River flow	3,814	4,538	4,746	5,040
Groundwater	221	301	351	385
Reservoir	1,213[a]	4,593[b]	4,599[c]	5,932[d]
Final balance	157	1,687	972	1,441

[a]Water supply from Soyanggang Reservoir.
[b]Water supply from Choongju Reservoir (3,380 million m³).
[c]Includes Kwangdong Reservoir (6 million m³).
[d]Includes Hongcheon Reservoir (1,333 million m³).

Source: Ministry of Construction 1980.

structed. Second, the share of agricultural demand in the Han River basin is less than 20 percent, compared to more than 40 percent in other river basins. This is because most large farmlands are located in the southern part of the Korean peninsula. Third, water supply in the Han River basin depends heavily on multipurpose dams, which will contribute more than one-half of total water supply by 2001. Table 7.3 lists seven multipurpose dams in Korea with a total annual water supply capacity of 8,300 million m³; Soyanggang and Choongju dams on the Han River account for 55.3 percent of this figure.

Within the Han River system, the South Han River basin uses more than 50 percent of the total agricultural water, and the Lower Han River basin consumes another 30 percent. More than 80 percent of residential and industrial water demand is concentrated on the Lower Han River, which flows through Seoul. To discourage further concentration and to satisfy water demand, the Revised Second Comprehensive National Land Development Plan (1987–1991) calls for relocation of population and industries in the Seoul Metropolitan Region. The region was classified into five areas: Seoul and its suburbs are classified as a "Move-Encouraging Area," where the establishment of new resident-increasing facilities such as universities are prohibited, but the transfer of existing facilities is encouraged. The plan classifies the Upper Han River basin as "Nature-Preservation Area," where the establishment of environmentally

TABLE 7.3   Multipurpose Dams in Korea

Dam	Feature Height (m)	Feature Length (m)	Total Capacity (million m^3)	Benefit Flood Control (million m^3)	Benefit Water Supply (million m^3/year)	Benefit Electricity (million kwh/year)
Total			9,040	1,530	8,300	1,808
Soyanggang	123	530	2,900	500	1,213	353
Andong	83	612	1,248	110	926	158
Namgang	21	975	136	43	134	43
Sumjingang	64	344	466	27	350	160
Daechung	72	495	1,490	250	1,649	250
Choongju	97.5	447	2,750	600	3,380	844
Nakdongang Estuary	—	2,400	50	—	648	—

Source: Ministry of Construction 1988.

damaging industries is prohibited and various restrictions are put on the activities of existing businesses, to preserve the water source of the Seoul Metropolitan Region.

Water Supply Through Areawide Waterworks

Increasing demand for water in and around Seoul can no longer be satisfied by the waterworks of local governments. Since the rivers near big cities are polluted and new sources of clean water are hard to find, each city seeks water from more distant rivers. However, such developments could be costly and inefficient, as they might result in overlapping investments and externality problems. It is more efficient for a specific water supply authority to construct areawide waterworks and to distribute the raw water to nearby cities and towns.

The waterworks system in Korea is dualized into areawide waterworks and local waterworks. Areawide waterworks are constructed by the Ministry of Construction and managed by the Korea Water Resources Corporation, a government corporation. Local waterworks are constructed and managed by local governments. Areawide waterworks supply water to about one-half of total benefited population in Korea, mostly city residents.

The Korea Water Resources Corporation sells raw water and treated water to the water authorities of local governments through areawide waterworks. The water fee charged by the corporation includes a basic-use fee and a measured-use fee. The basic-use fee is set by dividing total fixed cost by initially contracted quantity of water. The measured-use fee is equal to the average variable cost, which is total variable cost divided by total water actually supplied. Thus, the rate structure of areawide waterworks is close to the average cost-pricing scheme, which sets price just high enough to cover supply cost, including a normal rate of return on invested capital. In addition to the basic- and measured-use fees, an excess-use fee can be added when necessary for efficient distribution of water. Water fee revisions are made by the Ministry of Construction, upon the request of the Korea Water Resources Corporation, in consultation with the Economic Planning Board. Currently, the Korea Water Resources Corporation charges 38.03 won/m³ for raw water and 77.14 won/m³ for treated water uniformly to all users of areawide waterworks.

As of 1988, total capacity of areawide waterworks in Korea was 3.05 million m³/day, with an additional capacity of 2.06 million m³/day under construction. The Metropolitan Region Areawide Waterworks project was started in 1973, and stages 1 and 2 were completed by 1981. Stages 1 and 2 have a total capacity of 2.6 million m³/day, supplying good quality water from Paldang Reservoir to eight cities including Seoul and Inchon in the Seoul Metropolitan Region. Stage 3 started in 1984 and was completed in 1988. It supplies 1.3 million m³/day to twenty-five cities. Stage 4 started in 1989 to supply an additional 1.5 million m³/day to twenty-nine cities.

Table 7.4 shows the raw water sources for Seoul Metropolitan Region. In 1988 Seoul received 66 percent of its raw water from the Lower Han River and bought 34 percent from Paldang Reservoir through the Metropolitan Region Areawide Waterworks. In Inchon, 76 percent of raw water came from Paldang Reservoir, and 24 percent was pumped from the Lower Han River. In Kyunggi Province, 75 percent of raw water came mainly from Paldang Reservoir, 16.2 percent from the riverbed in the Han River tributaries, and only 7.5 percent from the surface flow of the Han River or its tributaries. The main flow of the Lower Han River still makes up more than one-half of the total raw water supply to Seoul and Inchon. However, cities and towns in Kyunggi Province receive and use only a small amount of raw water from the surface flow, since the Han River is far away and its tributaries are too polluted. All raw water transmitted from Paldang Reservoir to Seoul, Inchon, and Kyunggi Province is supplied by the Korea Water Resources Corporation through the Metropolitan Region Areawide Waterworks.

City Water Supply in the Seoul Metropolitan Region

According to the Sixth Five-Year Economic and Social Development Plan (1987–1991), the nation is divided into four extensive economic regions along four large rivers in South Korea. One of them is the Seoul Metropolitan Region, which receives water from the Han River. This

TABLE 7.4 Seoul Metropolitan Region Waterworks Capacity by Raw Water Source (1988) (1000 m³/day)

Area	Water Plants	Capacity	River Flow	Paldang Riverbed	Paldang Reservoir	Ground-water
Seoul	9	4,970 (100%)	3,290 (66%)		1,680 (34%)	
Inchon	2	1,060 (100%)	250 (24%)		810 (76%)	
Kyunggi Province	54	1,331 (100%)	101 (7.5%)	216 (16.2%)	993 (75%)	21 (1.3%)
Total	65	7,361 (100%)	3,641 (49.5%)	216 (3%)	3,483 (47.3%)	21 (0.2%)

Source: Ministry of Construction 1987b.

region includes Seoul Special City, Inchon Directly Governed City, and Kyunggi Province. The Seoul Metropolitan Region accounts for 39 percent of the South Korean population and produces 40 percent of total national output. Thus, the importance of the Han River to the Korean people and economy is enormous.

The concentration of population and improved living standards brought about a rapid increase in the demand for city water in Korea. In 1986, 68 percent of the people in South Korea received city water, with a daily per capita supply of 295 liters. Water services in Seoul are far better than the national average (see Table 7.5). In 1986, 98 percent of Seoul residents received city water, with a daily per capita supply of 390 liters. Ninety-five percent of Inchon residents received an average daily supply of 329 ℓ/capita. However, only 72 percent of the people living in Kyunggi Province were serviced, with an average daily supply of 214 liters, which is only 55 percent of the level of Seoul. This inequality of water supply between big cities and small-to-medium towns is one of the problems that must be solved.

Another problem is the high loss ratio that occurs during city water transmission. Transmission losses range from 23 to 47 percent in the Seoul Metropolitan Region. The high delivery losses are primarily caused by aged waterworks pipes. In the network of 17,185 km of pipe, only 65.1 percent is less than 10 years old, 27.6 percent is 11 to 20 years old, and 7.3 percent is more than 20 years old. Besides causing water losses, old pipes increase the risk of water contamination during its delivery to the final users.

TABLE 7.5   City Water Supply in the Seoul Metropolitan Region (1986)

Area	Total Population (A)	Benefited Population (B)	(B) / (A) (%)	Daily Total (1000 m³)	Daily Per Capita (ℓ)
Seoul	9,798,542	9,573,167	98	3,729	390
Inchon	1,441,131	1,376,280	95	454	329
Kyunggi Province	3,965,854	2,848,700	72	610	214

Source: Ministry of Construction 1987a.

Since local waterworks are operated by separate local governments, rate structures for city water vary throughout the country. Also, different rates are charged for different usage classes. City water fees for industrial and commercial uses are generally higher than for domestic use fees. Since a uniform or progressive rate structure, combined with basic fees, is employed within the same usage class, large water users pay relatively more than small users. Seoul uses nine categories, including domestic, commercial, industrial, and public. Table 7.6 shows the water rate structure for domestic water use in Seoul and Kyunggi Province. Basically, the city water fee is the sum of the basic fee and an additional progressive fee. The basic fee for Seoul is 600 won for the first 10 tons, and an additional charge based on consumption. Seoul's domestic water fee is one of the lowest in the country; some cities charge about five times as much.

Commercial water users in Seoul, including manufacturing and service industries, pay 4.6 times more for the basic fees and 5 times more for the additional fees than do domestic users. In this category the basic fee is 8,270 won up to the first 30 tons, and the additional fees are 410 won/ton for 31 to 200 tons, 550 won/ton for 201 to 300 tons, and 700 won/ton over 300 tons of city water used.

TABLE 7.6   City Water Rate Structures for Domestic Customers in Seoul and Kyunggi Province

Region	Basic Charge (won/10 tons)	Additional Charge (won/ton) 11–30 t	31–50 t	Over 51 t
Seoul	600	79	174	230
Kyunggi Province	560–1,190	80–150	100–190	140–270

Source: Korea Research Institute for Human Settlements 1987.

TABLE 7.7   Quality of Lower Han River at Water Intake Stations (1986) (mg/ℓ)

Intake Station	BOD	COD	DO
Paldang	1.4	2.1	10.9
Kuui	1.7	3.3	10.1
Ttukdo	2.5	4.0	9.9
Bokwangdong	3.0	4.1	9.5
Noryangjin	4.1	4.7	8.5
Sonyu	4.2	5.3	7.6

Source: Office of Environment 1987.

Deteriorating Water Quality

Han River Pollution and Restoration Efforts

As mentioned, the discharge of urban and industrial wastes exceeds the river's self-purification capacity and has polluted the Lower Han River. Table 7.7 shows the water quality of the Lower Han in 1986. One can see that the water quality worsens as the river flows downstream. Of the six intake stations along the Lower Han River, Paldang, Kuui, and Ttukdo passed the water quality standard of class 2 (suitable for swimming) as specified in the Environmental Preservation Law. All other stations passed the class 3 standard (industrial grade), which was set as a minimum for water to waterworks (see Figure 7.2). The trend of 1980–1986 shown in Figure 7.2 indicates that the quality of the Lower Han River has improved since 1985. The completion of the Han River Comprehensive Development Project is credited for improving the quality of water flowing from the tributaries along the Lower Han River.

The Han River Comprehensive Development Project was started in 1982 for improving the quality and appearance of the Han River before the 1988 Seoul Olympics. This scheme included three basic and two accompanying projects. The basic projects were to maintain the average depth of waterway at 2.5 m, to provide high-water parks along the river, and to expand the freeway from Kimpo Airport to Olympic Town. The two accompanying projects were the major contributors to quality improvement of the Han River. First, a sewer pipe system, which runs 53.8 km along the river, was constructed to collect domestic and industrial wastewaters from the city and to deliver them to sewage treatment plants, thus preventing untreated waters from flowing directly into the river. Second, three new sewage treatment plants were constructed to treat the polluted waters from three tributaries—the Jungrangchun,

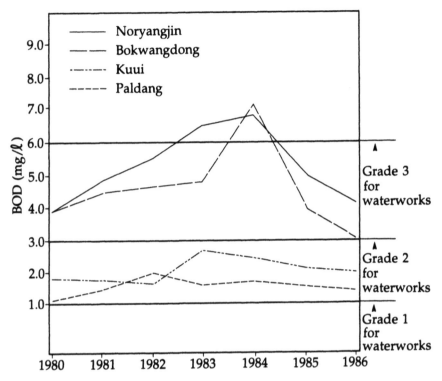

FIGURE 7.2 Water quality trend in the Lower Han River (*Source:* Office of Environment 1987).

Tanchun, and Anyangchun rivers (see Figure 7.1). After this project was completed, the sewage treatment capacity of Seoul increased from 360,000 to 3,060,000 tons/day.

Although the Han River Comprehensive Development Project improved the Lower Han River water quality, upstream water quality has continued to deteriorate, and the level of sewage treatment is not yet satisfactory. A recent survey conducted by the Korea Water Resources Corporation revealed that BOD levels in Paldang Reservoir increased from 1.4 mg/ℓ in 1986 to 2.4 mg/ℓ in 1988. This is a serious threat to the residents of the Seoul Metropolitan Region, because the reservoir supplies 34 percent of raw water to Seoul and nearly 75 percent to other cities and towns in the region. The main cause of pollution is inflow of wastewater from residential areas, industrial plants, and animal farms upstream from Paldang Reservoir.

The Ministry of Construction states that 19 million tons of sewage is produced daily throughout the nation, of which only 26 percent is

TABLE 7.8   Water Treatment Capacities in Seoul City (1988) (1000 m³/day)

Plant	Capacity by Raw Water Source		Total
	Paldang Reservoir	*Lower Han River*	
Paldang	1,000	—	1,000
Amsa	—	1,000	1,000
Kuui	—	1,130	1,130
Ttukdo	—	500	500
Bokwangdong	—	300	300
Noryangjin	160	140	300
Sonyu	180	220	400
Yongdungpo	240	—	240
Kimpo	100	—	100
Total	1,680 (34%)	3,290 (66%)	4,970 (100%)

Source: Ministry of Construction 1987a.

treated and the rest is discharged into rivers untreated. Even Seoul did not have any sewage treatment plants until 1979, and the present capacity handles only 82.3 percent of sewage generated. Also, the treatment level is insufficient. The Anyang and Nanji plants treat badly polluted wastewaters only partially, thus discharging water that is still polluted far beyond the wastewater effluent standards.

Seoul started to charge sewerage user fees in 1983, based on a "beneficiary pays principle." It is charged to all city water users, in addition to the city water fee. Users of groundwater and surface and industrial water are also charged the sewerage user fee separately. The fee is charged according to six categories of usage, and a progressive rate structure is employed to encourage water conservation.

Han River Pollution and Increasing Water Cost

Nine water purification plants along the Lower Han River supply water to Seoul. Of the total capacity of 4.97 million m³, 66 percent is taken from the Lower Han River and 34 percent from the Paldang Reservoir (see Table 7.8). Seoul takes 660,000 m³/day from the Lower Han River below Jungrangchun, where the water is most severely polluted.

There is a positive correlation between the water quality and water treatment costs. One study compared the unit operating costs of two water purification plants located on the Lower Han River for 9 months in 1982. The Paldang plant receives water from Paldang Reservoir, and the Bokwangdong plant takes raw water from the Lower Han River (see Figure 7.1). The results indicate that the unit chemical cost of the Bokwang-

TABLE 7.9 Operating Costs at Paldang and Bokwangdong Water Purification Plants (won/m³)

Item	Paldang	Bokwangdong
Chemical cost	0.957	7.411
Electricity	0.277	0.277
Labor cost	1.773	2.605
Total	3.007	10.293

Source: Office of Environment 1983.

dong water purification plant is more than seven times as high as at the Paldang plant (see Table 7.9). Unit energy and labor costs have no direct relation to the quality of raw water. Chemical treatment costs were 72 percent of total operating costs in the Bokwangdong plant, and only 32 percent of costs at the Paldang plant. Unit operating costs of the former were more than three times higher than the latter.

Seoul has three options now to secure raw water for municipal waterworks: (1) continue taking water from the polluted area and pay higher treatment costs; (2) construct new water intake facilities upstream where water quality is better; or (3) purchase more from the areawide waterworks, which supplies good quality raw water. In any case, Seoul must pay more to get the same quantity and quality of water than before, because of river pollution.

Water Organizations and Institutional Arrangements

Organizations and Legal System

Water resource management can be classified into four categories: water resource development, river management, water resource use, and water quality control. Since several government organizations are involved in water resource management at different stages, overlapping investments and conflicts of interest among them are possible, causing efficiency losses and difficulties in achieving welfare maximization.

First, in water resource development, the Ministry of Construction is responsible for constructing multipurpose dams. The Ministry of Agriculture, Forestry, and Fishery has jurisdiction over the construction of irrigation dams, and the Ministry of Energy and Resources oversees the construction of hydropower plants. Laws related to these activities are the Specific Multipurpose Dam Act, the River Act, the Electricity Business Act, and the Public Enterprise Act. Groundwater is developed by specific users such as local governments for domestic use, private persons for commercial or industrial uses, and the Agricultural Development Corporation for agricultural use.

Second, river management is dualized in Korea. "Directly controlled" rivers are managed by the Ministry of Construction, whereas local rivers are managed by local governments such as cities or provinces based on the River Act.

Third, the national water use plan is the responsibility of the Ministry of Construction. The Korea Water Resources Corporation constructs and operates areawide waterworks and industrial waterways, and local governments oversee local waterworks and supply city water to the final users. The Economic Planning Board (EPB) approves proposed waterworks plans following investment priorities, and evaluates and approves waterworks fees requested by the Ministry of Construction. City water fees are calculated by respective local governments based on total cost, including production cost and the rate of return on invested capital, which is about 7 percent in Korea. Local governments consult with the EPB when changing city water fees based on city ordinance, because the EPB supervises public utility fees. Management of water use and waterworks are provided for in the Waterworks Act.

Fourth, water quality control is mainly supervised by the Office of Environment, which was established in 1980 and expanded to the ministerial level in 1990. Although overall environmental policy is formulated by the Ministry of Environment based on the Environmental Preservation Law, several organizations are involved at the implementation stage. Based on the Sewerage Act, construction and management of sewerage systems are the responsibility of the Ministry of Construction. The Ministry builds sewerage systems in industrial complexes directly and supports construction costs to local governments—30 to 50 percent of construction costs for large cities and 70 percent for small-to-medium cities. However, Seoul constructs and manages its sewerage system independently, and the Ministry of Environment supervises nightsoil and wastewater treatment plants.

The water quality standards set by the Environmental Preservation Law state that the biochemical oxygen demand (BOD) should be below 6 mg/ℓ to qualify as raw water for waterworks, and below 10 mg/ℓ for industrial use (see Table 7.10). No section of the Lower Han River satisfies the water quality standard for grade 1 raw water for waterworks. Some of the water treatment plants barely pass the standard for grade 3 raw water sources, a warning that the river flow might not be available for future use in the waterworks. BOD levels of some tributaries in Seoul, such as the Chunggyechun and Anyangchun, exceeded 100 mg/ℓ before they improved slightly in 1985.

TABLE 7.10   Water Quality Standards (Rivers)

				Standard		
Class	Purpose	Hydrogen Ion (pH)	BOD (mg/ℓ)	SS (mg/ℓ)	DO (mg/ℓ)	Coliform (MPN[a]/100 mℓ)
I	Water supply (grade 1); Natural environment preservation	6.5–8.5	<1	<25	>7.5	<50
II	Water supply (grade 2); Fishery (grade 1); Swimming	6.5–8.5	<3	<25	>5	<1,000
III	Water supply (grade 3); Fishery (grade 2); Industrial use (grade 1)	6.5–8.5	<6	<25	>5	<5,000
IV	Industrial use (grade 2); Agricultural use	6.0–8.5	<8	<100	>2	—
V	Industrial use (grade 3); Living environment preservation	6.0–8.5	<10	No floating wastes	>2	—

[a]MPN = most probable number

Source: Ministry of Environment 1991.

Cost Allocation of Multipurpose Dams

The Korea Water Resources Corporation was established in 1967 based on the Water Resource Development Corporation Act to implement the activities specified in the Specific Multipurpose Dam Act and the Water Resource Development Act promulgated in 1966. Based on the Ten-Year Water Resource Development Plan initiated in 1966, the Soyanggang Multipurpose Dam was constructed for flood control, electricity generation, and water supply mainly for the Seoul Metropolitan Region. The Choongju Multipurpose Dam on the South Han River served the same purposes, plus supplying irrigation water. The two dams doubled the capacity of potable water supply from the Han River.

The Korea Water Resources Corporation assumed the responsibility for metropolitan waterworks facilities in 1981 from the Ministry of Construction and started to supply raw water, as well as purified water, to Seoul and other cities in the Seoul Metropolitan Region. To recover costs to continue multipurpose dam construction, the Water Resources Corporation allocates costs to the beneficiaries by the benefits they are expected to receive. Table 7.11 shows the cost allocation schemes for the two multipurpose dams on the Han River. More than one-half of the net

TABLE 7.11   Cost Allocation of Multipurpose Dams (million won)

Benefits	Soyanggang Dam		Choongju Dam	
Flood control	5,468	(19.0%)	93,107	(16.9%)
Domestic and industrial water	6,287	(21.8%)	79,501	(14.4%)
Irrigation	—		15,952	(2.9%)
Electric generation	17,083	(59.2%)	363,857	(65.8%)
Net construction cost	28,838	(100.0%)	552,417	(100.0%)
Unclassified	3,247		2,630	
Total construction cost	32,085		555,047	

Source: Korea Water Resources Corporation 1987.

construction cost of both dams is allocated to the Korea Electric Power Corporation. Domestic and industrial users bear 22 percent of the costs of Soyanggang Dam and 14 percent of Choongju Dam. The remaining construction costs are allocated to the beneficiaries of flood control and irrigation water.

Although the capacities of Soyanggang and Choongju dams are nearly the same (2,900 and 2,750 million m^3, respectively), the construction cost of Choongju Dam was seventeen times higher than that of Soyanggang Dam; the per unit construction cost was 11.1 won versus 201.8 won/m^3. Rapidly rising construction costs and lack of proper dam sites make it difficult to construct additional multipurpose dams on the Han River. As an alternative, construction of small dams or water transfer from other rivers can be considered. But eventually demand-side management through introduction of water-recycling systems, new water-pricing schemes, and transfer of water between sectors should be seriously considered in coping with water shortages in the Seoul Metropolitan Region.

Summary

To meet rapidly increasing demand for water in the Seoul Metropolitan Region, Soyanggang and Choongju multipurpose dams were constructed, and areawide waterworks were developed. This kind of supply-side management delayed water shortages in the region. However, rapidly rising construction costs and lack of suitable dam sites make it increasingly difficult to construct additional multipurpose dams on the Han River.

In addition to the absolute limit in the water supply, potable water is increasingly hard to get. The quality of the Lower Han River has been deteriorating because of urban and industrial wastes from the Seoul Metropolitan Region, and because the Paldang Reservoir, which supplies 34 percent of raw water to Seoul and 75 percent to other areas in Kyunggi Province, is being polluted by the inflow of wastewaters from farms and towns upstream.

The implication of diminishing potable water supply in the Seoul Metropolitan Region is increasing conflicts among water users in the Lower Han River basin. Conventional water-supply management is still effective and necessary, but not sufficient. To cope with the upcoming water shortage problems, a more efficient way of allocating scarce water resources should be developed. Since different government agencies are involved in various stages of water resource management, overlapping investments or conflicts of interest sometimes cause efficiency loss and government failure. A comprehensive water resource management system, which considers water development, water use, and water quality preservation at the same time, should be developed and existing institutions should be properly reorganized. Eventually demand-side management through water conservation, introduction of marginal cost pricing, and transfer of water use between sectors should be adopted.

References

(All references are government documents in the Korean language and printed by the Korean government in Seoul, Korea.)

Korea Research Institute for Human Settlements. 1987. *Waterworks Rate Structure Rationalization Plan.*
Korea Water Resources Corporation. 1987. *Long-Term Development Strategy to the Year 2000.*
Ministry of Construction. 1980. *Long-Term Comprehensive Water Resource Development Base Plan (1981–2001).*
Ministry of Construction. 1987a. *Revised Plan of the Second Comprehensive National Land Development Plan (1987–1991).*
Ministry of Construction. 1987b. *Waterworks.*
Ministry of Construction. 1988. *Annual Report on National Land Use.*
Ministry of Environment. 1991. *Korea Environmental Yearbook 1991.*
Office of Environment. 1983. *Comprehensive Han River Basin Environmental Preservation Plan: Report on Water Quality.* Vols. 1–4.
Office of Environment. 1987. *Environmental Pollution Survey (1980–1986).*

8

Water Use Conflicts in Bangkok Metropolitan Region, Thailand

Ruangdej Srivardhana

The Bangkok Metropolitan Region (BMR), which consists of the Bangkok Metropolis and parts of the five adjacent provinces of Nonthaburi, Samut Prakan, Nakhon Pathom, Pathum Thani, and Samut Sakhon, occupies an area of approximately 6,000 km² in the lower part of central Thailand. The city of Bangkok itself covers 1,570 km² and is located at the inner zone of the BMR. The BMR is established for the planning and management of public infrastructure such as roads, highways, electricity, and the water supply of the six neighboring cities. These cities are rapidly expanding toward one another in converging with Bangkok into one megacity. Despite this growing interconnection, however, each province is a separate unit politically and administratively under the central government's control.

Geographically, the BMR is located at 13°30'–13°55' N and 100°25'–100°45' E. It has a subtropical humid climate, with average annual rainfall of 1,200 mm/year. The region is transected by the Chao Phraya River which is, up to now, the only surface water resource for metropolitan water supply. The Bangkok Metropolis is located about 25 km from the mouth of the Chao Phraya River. After passing through Bangkok Metropolis, the river flows through the province of Samut Prakan before discharging into the Gulf of Thailand (see Figure 8.1). The entire BMR lies on low, flat land with an altitude above mean sea level of 0.1 to 2.0 m, averaging 1.5 m.

The region is known as the nation's development center, where most of the nation's important economic activities such as trade, industry, and other services are concentrated. In addition, as the nation's capital, the Bangkok Metropolis hosts all important central government offices, and it is here where most national decision-making takes place. The region's

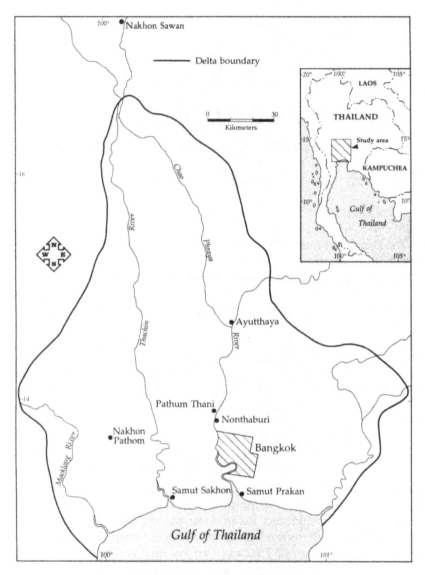

FIGURE 8.1 Bangkok and vicinity.

high status is made clear because about one-half of the national economic growth takes place within this particular region. In 1986, about 15 percent of the country's total population resided here (Table 8.1). Population is expected to increase to 18 percent of the national total in 2001, reaching 11.5 million. Although the BMR will become more urbanized and densely populated with development, some parts will remain rural. The National

TABLE 8.1   Past and Projected Population in the BMR

Province	Population (thousand)			Growth Rate (% / year)	
	1986	*1991*[a]	*2001*[a]	*1986–91*	*1986–2001*
Bangkok	5,468.9	6,477	7,850	2.3	2.1
Nakhon Pathom	617.6	672	796	1.8	1.7
Nonthaburi	525.5	556	782	3.3	3.4
Pathum Thani	402.1	478	681	3.3	3.5
Samut Prakan	699.6	739	1,002	3.4	3.2
Samut Sakhon	327.7	331	430	2.4	2.6
Total BMR	8,041.4	9,253	11,541	2.5	2.3
Thailand[a]	52,969.2	57,196	65,138	1.7	1.4
BMR/ Thailand (%)	15.18	16.18	17.72		

[a]Data drawn from the "medium" projection of the Working Group on Population Projections (comprising NESDB, the National Statistics Office, and the Institute of Population Studies, Chulalongkorn University).

Source: Adapted from TDRI 1988.

Economic and Social Development Board (NESDB) has predicted an increase of the BMR urban share of population from 83 percent (7.7 million) to 86 percent (9.9 million) between 1991 and 2001 (Table 8.2). These figures imply a slight increase in the absolute number of the rural population, from 1.6 to 1.7 million.

TABLE 8.2   Projected Urban Population in the BMR (thousand; % of urbanized population)

Province	1991		2001	
Bangkok	6,477	(100.0)	7,850	(100.0)
Nakhon Pathom	151	(22.5)	247	(31.1)
Nonthaburi	248	(44.6)	460	(58.9)
Pathum Thani	233	(48.8)	427	(62.7)
Samut Prakan	415	(56.2)	655	(65.3)
Samut Sakhon	139	(42.0)	228	(53.0)
Total BMR	7,663	(82.8)	9,867	(85.5)
Whole Kingdom	17,159	(30.0)	22,863	(35.1)

Source: NESDB 1986.

Since water is a major production input and an essential element of living standards, it is important to know its status, both in terms of quantity and quality. Inadequacy of water can hamper economic development and the quality of life of the BMR people. This chapter discusses water use conflicts within the BMR and the legal framework that can be used to solve or reduce such conflicts. It first surveys the BMR water resource availability and the potential for both surface water and groundwater. It then elaborates on the demand side, with a future demand estimation. The demand-and-supply analysis then follows. Further, the degradation of the lower Chao Phraya River by human activities will be presented, as water quality directly relates to the amount of usable water downstream that affects those who live by the river.

Last is a discussion on legal aspects of water management in Thailand. The country's existing water laws and regulations must provide a foundation for water use and allocation, and hence have an effect on the BMR water resources situation.

The BMR's Water Resources

Surface Source

At present the only surface water source for the BMR is the Chao Phraya River, although many rivers in nearby basins are good potential sources. The Chao Phraya River originates from four tributaries (i.e., the Ping, the Wang, the Yom, and the Nan) in the country's upper watersheds, 600–700 km north of the BMR. The Wang Tributary merges into the Ping, and the Yom into the Nan rivers, respectively. These two rivers then converge in the province of Nakhon Sawan in central Thailand to form the Chao Phraya River. Above the confluence that forms the Chao Phraya River, there are two impounding dams, the Bhumiphol and the Sirikit, on the Ping and the Nan rivers, respectively (Figure 8.2). The major benefits of these two dams, which are multipurpose in nature, include electricity generation and irrigation.

The 360-km Chao Phraya River flows by and through many towns and cities, of which the BMR is the greatest, before it flows into the Gulf of Thailand. Water in the Chao Phraya is diverted by the Chao Phraya Barrage at Chainat, 160 km from the estuary. The Bhumiphol and Sirikit dams release approximately 145 m³/sec to the barrage, which in turn allocates it as follows:

1. *Urban water supply to the BMR (25 m³/sec).* This water, conveyed by canal from Samlae, 90 km from the river's mouth, to treatment plants of the Metropolitan Waterworks Authority (MWA), constitutes a major portion of the BMR total water supply. At present approximately 2 mil-

FIGURE 8.2 The Chao Phraya River basin, its upper tributaries, and neighboring river basins.

lion m³/day is withdrawn for this purpose. The rest of the BMR's water supply comes from groundwater.

2. *Irrigation* (25 *m³/sec*). An equivalent amount of water is currently diverted for irrigating approximately 7.5 million *rai* (1.2 million ha) on both sides of the Chao Phraya below the barrage. Unfortunately, the water is only enough to irrigate dry season (second crop) rice on 2.5 million *rai*. Although the farmers would like to grow dry season rice on all irrigation land, they must alternate and do so only once in 3 years.

3. *Prevention of seawater intrusion into the Thachin (45 m³/sec)*. This water is diverted through a natural channel to the Thachin River, which runs parallel to the Chao Phraya (Figure 8.1), to prevent saline water intrusion.

4. *Prevention of seawater intrusion into the Chao Phraya (50 m³/sec)*. A slightly greater amount is devoted to flushing out seawater, which would otherwise encroach into the inner stretches of the Chao Phraya. The main purpose of this is to protect agricultural production in the BMR, notably in Nonthaburi Province, located just north of Bangkok Metropolis. Nonthaburi is famous for its production of fruits such as durian, santol, banana, and citrus, which generate more than 500 million baht (US$20 million) in income each year. Most of the orchards are along the Chao Phraya and its tributaries and could be destroyed if seawater penetrates 40–50 km upstream.

The present surface allocations are deemed just adequate for present urban water supply but far from enough for dry rice farming. A conflict already exists within agriculture, between upstream rice growers and downstream fruit producers. If the 50 m³/sec currently used to protect orchards from salinity were diverted to irrigation, rice could be grown on all irrigable land during the dry season.

With shortages looming in the urban areas, the BMR is considering diverting the 45 m³/sec, which is now discharged into the Thachin, to prevent salinity in that river as a possible urban water source. This could be retained in the Chao Phraya, provided that the same flow into the Thachin is made available from elsewhere. Fortunately, the adjacent Maeklong River Basin, west of the Thachin, is endowed with water so plentiful that there is a surplus after fulfilling all demands within the basin. Furthermore, a natural channel already exists that could be used to convey water to the Thachin. Thai authorities see this diversion as an effective, potential means of increasing surface water resource to the BMR, from 25 to 70 m³/sec, without requiring major structural investment or infringement on other existing water uses.

Groundwater Sources

Withdrawal of groundwater in small amounts within the BMR has been reported since the 1920s. Early withdrawals were from shallow wells since the groundwater level was still high then. Large-scale water withdrawals did not take place until Thailand inaugurated its first National Economic and Social Development Plan in 1961. Because of a rapid increase in water demand resulting from Bangkok's expansion, the MWA tapped groundwater for local uses in areas where the central water supply service was still not available. In addition, those in the private sector, including households and industries, who could not wait for the MWA's water supply, tapped groundwater for their own uses. In 1963 the MWA, which was the largest groundwater user at the time, pumped approximately 300,000 m³/day for its water supply. Statistics of private wells were not available before 1977 when the Groundwater Act was promulgated. Comprehensive statistics of groundwater withdrawal were first available in 1982, when approximately 1.4 million m³/day was extracted. Of this total, about 0.96 million m³/day was withdrawn by private wells, the remainder by the MWA.

These estimates were based on reported statistics which, because of inadequate monitoring and enforcement, tend to be underestimated. The actual amount of groundwater withdrawal could be much higher.

Nonetheless, even the underestimated figures indicate the importance of groundwater in supplying the BMR's water. They also show a significant overdraft of this natural resource. Studies have indicated a conservative estimate of 0.6 to 0.8 million m³/day was withdrawn in excess of the basin's recharge capacity (e.g., Sharma 1986:11–15). This directly affected the piezometric surface (artesian water table), which at one time was to the ground level. In 1959 the deepest level recorded was 12 m below ground surface in central Bangkok. After 1966–1967, the level sank by 30 m in the same area. By 1981 the lowest piezometric level was about 52 m below ground surface. The Asian Institute of Technology (1982) found that the average decline of the piezometric surface was 2.5 m/year during 1972–1982.

Excessive groundwater withdrawal is a major known cause of land subsidence. For the BMR, land subsidence ranged from 5 cm to more than 10 cm/year throughout the region. The most critical area was eastern Bangkok, where the decline was more than 10 cm/year. This was the area where the city's most rapid expansion took place, and the MWA could not respond adequately with alternative supplies to the rising water demand. The real needs of the BMR residents and economic activities led to excessive groundwater withdrawal and the land subsidence problem.

Land subsidence is evidenced by sinking of walkways, building floors, and streets, some of which become flooded during heavy rain. The costs to society are quite substantial. Sinking structures need repair and maintenance or even reconstruction. Flooded streets also cost society in terms of clean-up time. To alleviate the land subsidence problem, overdraft of groundwater must be halted. As mentioned earlier, pumping must be cut back by at least one-half, or 0.6 to 0.8 million m³/day, to allow for a sustainable yield (pumping-recharge balance). By adding 70 m³/sec of surface water from the Chao Phraya and Thachin rivers, 6.04 million m³/day of raw water could be made available to the BMR without requiring drastic structural improvement.

BMR Water Requirements

Since the BMR will continue to grow, its demand for water is expected to increase continuously. It is not possible to be precise about water requirements of each sector, except for the water supply, since there is not enough information.

The BMR's agricultural area will be reduced as urbanization continues and as more area is converted to housing and industrial uses. Agricultural demand for water in the BMR is defined as that amount which could prevent salinity from reaching the gardens and orchards. The 50 m³/sec of freshwater now released for this purpose from the Chao Phraya Barrage has also helped dilute pollution in the Chao Phraya River.

The water requirements of the BMR's rapidly expanding industrial sector are hard to estimate or project. The revised Master Plan of 1984 estimated the industrial groundwater withdrawal to be 610,000 m³/day, or 63 percent of total private abstractions, for April 1982 (see Sharma 1986:28). As mentioned, this is a rough estimate because of the inadequate monitoring and record keeping that accompany unregulated private development. Another problem is that officials of the MWA are reluctant to release data for fear that the information could be used against the agency.

If, as expected, the BMR's manufacturing sector continues to grow rapidly, industrial demand for water is likely to increase commensurately, both perhaps at 10 percent or more per year. This demand cannot be fulfilled from further depletion of the aquifer.

In 1983, the Cabinet passed a resolution to halt pumping of groundwater in the critical inner zones of the BMR. The MWA was directed to cease its own groundwater pumping in these zones by 1987, after which private withdrawals were to be phased out through 1998 (Sharma 1986:51). The MWA was allowed to abstract groundwater only in the outer urban communities that were not connected to the main water supply system.

This strategy places the burden for industrial water supply on the MWA. But the MWA will face increasing difficulty in acquiring freshwater for the BMR's urban water supply, as direct withdrawal from the Chao Phraya is no longer possible because of poor water quality.

Domestic Water Supply

A high percentage of households in the Bangkok Metropolis, Nonthaburi, and Samut Prakan have tap water. The MWA Act requires the MWA to supply water to these three provinces. Other provinces within the BMR are still outside the MWA water service area. Although useful, the MWA's statistics are inadequate for analyzing domestic water demand because statistics for domestic and industrial use are not recorded separately. Hence, the analysis that follows is suggestive rather than definitive.

The three provinces of Bangkok, Nonthaburi, and Samut Prakan had a total population of 6.9 million in 1985. Their population is expected to increase to 9.8 million in 2000.

To deal with a rapid increase in domestic demand, the MWA commissioned Camp, Dresser, and McKee to prepare a master plan. To cope with changing conditions, the master plan was extensively modified in 1984 by the Nihon Suido Consultants Co. and the Thai Engineering Consultants Co. (NS–TE). The MWA has since followed this modified plan for developing its water supply, although a review of the NS–TE work is now in process.

The modified Master Plan of 1984 considered many factors in its development plan. It provided for an expansion of the MWA service area from 486 km² in 1985 to 835 km² in the year 2000, to cover approximately 27 percent of the three provinces' area. It incorporated the 1983 Cabinet ban on groundwater pumping in critical zones. It envisaged that the number of people who must be serviced by the MWA central water supply system would increase from 4.3 million (85.3 percent of the total residents) in 1985 to 7.8 million (96.1 percent) in 2000. The modified master plan also stipulated that the unaccounted-for water (UFW) would decrease, because of improvements in management efficiency, from 43 percent of the overall distributed water in 1985 to 25 percent in the year 2000.

Taking these factors all together, the master plan estimated a requirement of approximately 5.2 million m³/day of raw water for the main MWA water supply in the year 2000. If, as intended, groundwater withdrawal is reduced to its sustainable yield of between 0.6 and 0.8 million m³/day, all 5.2 million m³/day will have to come from surface sources.

With only the present flow of 25 m³/sec, or 2.16 million m³/day, released from the Chao Phraya Barrage for the MWA water supply, the MWA will be short of surface raw water in the near future, especially when the requirement for water increases to 5.2 million m³/day in the year 2000. However, with a potential flow of 45 m³/sec, or 3.88 million m³/day, that the Chao Phraya River could save from diverting the Thachin River, the surface water resource is still adequate to meet the overall requirement. A portion of it, approximately 1 million m³/day, could be used for other purposes, especially irrigation in central Thailand which now has a lower priority than urban uses downstream.

The BMR's Surface Water Quality

The river's freshwater is withdrawn for the BMR's water supply at Samlae, 90 km from the gulf, where water is still free from the city's contamination. The water quality downstream, beginning in Bangkok Metropolis 60 km from the river's mouth, is normally poor because this stretch is used as a public sewerage by households and industries.

Water quality can be viewed from many perspectives. Among many well-known factors that directly affect water quality are heavy metals such as lead, cadmium, mercury; solid waste; coliform and other bacteria; and organic matter from households and industrial factories. Heavy metals were reportedly found in the water and fauna samples in the Chao Phraya's estuary and inner gulf. However, since the concentration of each metal was within an acceptable level, there was little concern about the problem. Acceptable levels were achieved because most of the large factories concentrated on the Chao Phraya River have complied with the effluent standards and have reduced heavy metals content before discharging wastewater into the river. Solid waste and coliform have not presented serious problems either, although they have occasionally attracted public attention.

Organic waste, when discharged into the water body, is decomposed by bacteria that uses oxygen in the decomposition process. Natural water of good quality contains about 7 mg of oxygen/ℓ. The decomposition process causes the dissolved oxygen (DO) in the water to decline, lowering water quality. Water with DO below 4 mg/ℓ has an adverse effect on some fish species such as carp. Water with DO below 2 mg/ℓ is not good for any fish. As DO approaches zero, a change of water color from clear to black and foul odors occur from the decomposition processes, causing damages to amenity-oriented uses such as tourism.

Organic Wasteload in the Chao Phraya River

The main cause of DO depletion, and hence water quality degradation, in the lower part of the Chao Phraya River is the disposal of excessive organic waste from the BMR's household and industrial establishments into the river. Since the Bangkok Metropolis and its satellite cities in the BMR do not have a public sewerage, all effluents, except that from the toilet and bath usually collected in cesspools, are discharged into canals (*klong*) before finally flowing into the Chao Phraya River. As mentioned, large factories play a relatively small role in deteriorating water quality since the government's environmental control agencies monitor their operations and stress compliance with environmental regulations and standards.

There have been many studies on organic wasteload, especially of industrial waste, discharged into the Chao Phraya River, with widely different estimates. Camp, Dresser, and McKee (1987) estimated that 145,000 kg/day of biological oxygen demand (BOD_5) was discharged into the river, 6,600 kg/day of which was from the industrial sector.

The National Environment Board (1983) reported that about 165,035 kg/day of BOD_5 was discharged by households from canals into the Chao Phraya River in 1979. This figure, together with 6,084 kg/day discharged directly into the river, totaled 171,119 kg/day of BOD_5 discharged from the household sector. The industrial sector was reported as discharging 95,644 kg/day. Therefore, from the NEB report, over 60 percent of the total wasteload of 266,763 kg/day was from the household sector, and the remainder was from manufacturers.

In a 1987 study, Panswad et al. indicated that the total nonindustrial BOD_5 load discharged into the Chao Phraya River was about 137,231 kg/day and that about 90 percent of this figure was from households (54.06 percent) and restaurants (36.91 percent). The rest was from sources such as dormitories, hotels, hospitals, and markets.

The Department of Industrial Works (1987) estimated that a BOD_5 load of 250,000 kg/day was discharged from the household sector and 8,265 kg/day from the industrial sector. The industrial BOD_5 figure was low because a treatment efficiency of 94.3 percent was applied to the overall 145,008 kg/day of the effluent generated.

With such divergent BOD_5 load values from various studies, the Thailand Development Research Institute (TDRI), in an attempt to encourage policy actions, reconciled the differences in a July 1988 conference. The participants, who were representatives of the agencies responsible for water quality, agreed that the overall BOD_5 load discharged into the Chao Phraya River in 1988 was approximately 183,634 kg/day. About 74.7 percent of this amount (137,175 kg/day) was from domestic sources and 25.3 percent from industrial sources (TDRI 1988).

Wastewater Treatment Dilemma

To improve the Chao Phraya's water quality, effluents discharged from the household sector must be reduced. Since it is neither possible nor efficient for each household to deal with its wastewater individually, a public sewerage to collect wastewater for mass treatment before releasing it into the river is a possible solution. The public sewerage and central treatment scheme have received attention since the early 1970s, when a system was proposed at a cost of approximately 30 billion baht. However, since this cost was too expensive for the government, nothing was done. Since Bangkok City has rapidly expanded during the last two decades and finally became the BMR, it now urgently needs an even larger sewerage and treatment system to cope with the increased amount of wastewater. The estimated cost of such a system for the entire BMR at present exceeds 100 billion baht (US$4 billion), about one-fifth of the current national budget. Despite concerns of the government and responsible agencies, construction is still not feasible because of the high cost.

As the BMR continues to prosper, water pollution in the lower stretch of the Chao Phraya River will continue to increase and hence create more damages to tourism and aesthetic uses. Conventional uses such as fishery, industrial, and domestic have already been deprived of river use by deteriorated water quality. But from an economic viewpoint, investment on the public sewerage and central wastewater treatment plant is not worth undertaking if the damage caused by water pollution does not exceed the cost of building such infrastructure. The Chao Phraya River's major service has been in receiving wastewater and other effluents in the BMR's development process. Without it, the society would have to pay for a garbage and effluent disposal system. The Chao Phraya River has saved the BMR more than 100 billion baht (US$4 billion).

From a political standpoint, a serious effort to deal with the Chao Phraya's pollution problem is not likely to happen unless there is pressure from interest groups such as those in tourism and other industries seriously affected by the pollution. Of course, the strength and willingness of different interest groups to act are closely linked to economic returns.

Summary of the BMR Water Use Conflicts

Upstream-Downstream Conflict

Spatial conflicts occur both upstream and downstream of the BMR. The Chiang Mai Valley in the upper watershed has been developing rapidly, and both urban and agricultural expansions require an increasing amount of water. Less water is therefore available for downstream uses, including irrigation in central Thailand, prevention of seawater intrusion into the inner stretch of the Chao Phraya River, and generation of electricity. This conflict reached a critical stage in 1993.

Upstream-downstream conflict can also be viewed from a water quality context. Deteriorated water quality in the lower stretch of the Chao Phraya River is caused by overusing the upper stretch as a public sewer. This indicates a clear conflict between industrial and household wastewater dischargers in the upper stretch and those who depend on the river for domestic, tourism, and industrial uses downstream, a conflict that is also sectoral in nature.

Sectoral Conflict

Many sectoral use conflicts take place in the BMR. Water saved for irrigation in the upper central plain cannot be used for water supply in the BMR. If the 25 m^3/sec for irrigation were given to the MWA, the raw water would be sufficient for the BMR population for a number of years. Instead, the MWA now faces a looming shortage of raw water. The 50 m^3/sec of water that is now released to prevent seawater from encroaching into the river's inner stretch could also be used by the MWA but at a cost of fruit production in Nonthaburi Province.

The Legal Framework

As Thailand continues to develop, more water will be required by different economic sectors, and conflicts among sectors will become more intensified. To allow for a smooth development, foreseeable conflicts must be reduced or, if possible, eliminated. Because water in Thailand is a natural resource for public use under government supervision, conflicts can be legally reconciled to a certain extent. It is therefore useful to understand the legal principles that govern water uses and management in Thailand, particularly those that could be used to reduce water use conflicts of the BMR.

Water-Related Laws

Thailand has no single water code. Instead, there are thirty-eight laws that directly or indirectly involve water resource utilization (see Annex). Those most related to water use conflicts of the BMR are discussed here.

The Civil and Commercial Code. This law is fundamental to all property rights and ownership. It does not include the detailed aspects of water, but gives important principles for all the water-related laws. The most important of these is the affirmation that since water is endowed by nature it should not belong to any person or party. Anyone in society has a right to water and cannot be prevented from using it. Water should be used for the maximum benefit of society and, as such, the Civil and Commercial Code recognizes water as being under government control or management, although individual landowners or tenants always have the right to use water on or beneath their land.

With the principle that everyone should be able to enjoy the benefits of water, Article 1339, Clause 2, of the code allows an upstream landowner to withdraw as much water as required from a source, as long as the water is then returned to the natural course for other uses downstream. Article 1355 prohibits excessive withdrawal of water because of potential harm to downstream users who depend on the same source of water.

As conflicts become more complex and intense, administrative solutions and simple regulations will no longer suffice, and the Civil and Commercial Code will become more important in resolving them.

The People's Irrigation Act, B.E. 2482. The Royal Irrigation Act (see below) gives the Royal Irrigation Department (RID) authority to take care of water within the proclaimed irrigation land that occupies approximately 20 percent of the country's arable land. Water use for cultivation and allocation elsewhere falls under the People's Irrigation Act of B.E. 2482 (A.D. 1939). The basic principles of this law remain in effect, although two amendments were made in 1980 and 1983. The law is the only one that can alleviate the upstream-downstream water use conflict, particularly between the Chiang Mai Valley and the Central Plain irrigation areas. But since water use conflicts have only recently become apparent to the Thai people, this law has been little used.

The law elaborates on the water rights specified by the Civil and Commercial Code and allows for only reasonable uses of water. Withdrawal of water for irrigation must be done in appropriate quantities and then returned to natural streams so that it can be used by others downstream.

Upstream-downstream conflict was mentioned earlier. Most of the Chiang Mai Valley in northern Thailand has experienced rapid development. The area is outside the RID jurisdiction and as such is subject to the People's Irrigation Act. As the Chiang Mai area develops further and brings additional land under irrigation, it will require more water. Unless the People's Irrigation Act is invoked, the upstream-downstream conflict will become more intensified since less water will be available downstream.

The law requires those who want to irrigate more than 200 *rai* (16 ha) to get a use permit from the authority. No one is allowed to retain water in excess of what they actually need for agriculture and to store water longer than a specified time, if doing so can prove harmful to others. If these measures are properly enforced, water will be used more efficiently in the Chiang Mai Valley upstream and more water will be available for downstream uses.

The Royal Irrigation Act, B.E. 2485. This act was established to improve irrigation water use. Before its enactment, Thai farmers used water rather freely. Only a part of the resource was used while a much larger portion flowed into the sea. The growth in Thailand's economy increased the demand for water and necessitated its more efficient use. The Royal Irrigation Act of B.E. 2485 (A.D. 1942) was promulgated for this purpose.

The law is based on the idea that water is provided freely by nature for everyone in the country. To make the best use of water, from a social viewpoint, ownership must be with the state which is presumed to operate in the people's best interest. The state has the prime authority to do anything with water, including allocation and conservation, for the maximum benefit of society.

The original purpose of the Royal Irrigation Act was to promote agricultural production, flood protection, and navigation. Two amendments, in 1964 and 1975, extended the scope of the law. The 1964 amendment authorizes the Royal Irrigation Department to take responsibility for water management within a designated irrigation area. It also brings some of the hydropower production and public health issues under the scope of this law. The 1975 amendment adds water for industrial uses. Hence, the scope of this act is now wide, covering various types of water use. Irrigation water can be used for agricultural production, water supply, or industrial needs.

The Central Plain of Thailand below the Chao Phraya Barrage is the country's rice bowl and a designated irrigation area. Massive structural investment has been made in this area to promote irrigation. Since the BMR lies within this irrigation area, the Royal Irrigation Act could be used to alleviate BMR water use conflicts.

The Royal Irrigation Department (RID) cannot by itself fully perform its duties as stated in the act. It has to collaborate with other water-related agencies such as the Electrical Generating Authority of Thailand (EGAT) and the BMR. Hence, allocation among different uses—such as irrigation, power generation, water supply, and prevention of seawater intrusion—is done by the National Water Resource Committee, chaired by one of the Deputy Prime Ministers. This committee consists of representatives from different water-related agencies including RID and EGAT. This does not end conflicts, but it is a way to allocate or limit water to competing uses.

Allocation of water in the dry season for irrigation in central Thailand is done through a national level committee, chaired by the Minister of Agriculture and Agricultural Co-operatives, with representatives from agencies responsible for crop production (i.e., departments of Agricultural Extension, Agriculture, Local Administration; the bank for Agriculture and Agricultural Co-operatives; and the RID). However, the actual proclamation that states the area to be supplied with irrigation water is made by the RID Director General under the Royal Irrigation Act.

Besides collaborating with other agencies in allocating water, the RID Director General or his representative is empowered by the Act to regulate the flow within a designated irrigation area. The RID is also responsible for maintenance works and elimination of wasteful uses. The irrigation officer is authorized to stop wasteful use, including storage, on any piece of land that deprives adjacent lands of water. Obviously, increasing water use efficiency means more water available for downstream uses.

Article 8 of the Royal Irrigation Act empowers the Minister of Agriculture and Agricultural Co-operatives to set charges for irrigation service on landowners or tenants within an irrigation area. The charge is very low. Within the irrigation area, a maximum charge of 5 baht/*rai* (US$1.25/ha at 1990 prices) per year is set. But in practice, because of political reasons, service charges have not been imposed on the farmers, although there have been many attempts to do so. The Minister may also levy a fee on other water users who withdraw water from irrigation facilities, no matter what their purpose for withdrawing water and wherever their lands are located. User charges are applied to industrial establishments and water supply authorities. Nonetheless, the charge for users outside the irrigation area, although progressive in nature, may not exceed 5 baht/m^3 (US$0.20/m^3).

Despite its history of disuse, Article 8 provides a legal basis for the authority to use demand management in coping with water conflict. An increase in the price of water, implemented effectively, would naturally discourage use, and more water will be released for other downstream uses.

The Groundwater Act, B.E. 2520. This law was enacted in 1977 (B.E. 2520) as a result of excessive groundwater use, especially in the BMR area. There was a strong anticipation that without control the groundwater resource would be exploited to the detriment of the resource and to society. The law permits a responsible authority to proclaim a groundwater zone where the groundwater resource is threatened. Any party within this proclaimed zone is required to have a permit for digging a well or for using the water obtained from a well. In addition, groundwater users within a proclaimed zone must pay for water at a price close to that of surface water supply available in that locality, to discourage groundwater uses. The law recognizes the importance of groundwater for those who reside outside the water supply service areas and exempts them from water charges if the water is used for domestic and agricultural purposes and does not exceed 25 m^3/day. They are still required to obtain permits for digging wells and using water, however.

Because of the land subsidence problem, the BMR was proclaimed a part of a groundwater zone, which includes Bangkok Metropolis and the provinces of Nonthaburi, Samut Prakan, Pathum Thani, Samut Sakhon, and Ayutthaya. Anyone who wants to use groundwater in this zone must have a water use permit issued by the Department of Mineral Resources.

In addition to use permits to discourage the use of groundwater, the law sets standards for drilling and digging for wells, and for recharge of surface water into the aquifer. It also establishes specific procedures for closing or abandoning a well and encourages conservation in groundwater use. All these requirements are to promote public health and the environment.

The law has several penalty clauses. For example, those who fail to follow the drilling and digging regulations are subject to a fine of up to 20,000 baht (US$800) (Article 37). Those who withdraw groundwater without a permit can be sentenced up to 6 months in jail and/or fined up to 20,000 baht (Article 38).

The Groundwater Act was followed 6 years later by a more restrictive 1983 Cabinet Resolution. As noted previously, this resolution prohibits unjustifiable groundwater use within the MWA service area. Hence, only those in the service area who can prove that the serviced water does not have the quality required by their needs may be allowed to use groundwater. The rest are obliged to use tap water.

To comply with the 1983 Cabinet Resolution, the MWA, which had been a major groundwater withdrawer, set a target of abandoning the groundwater resource by 1988. Although this target was not attained, the MWA significantly reduced its dependency on groundwater. It is hoped that in the near future it will reach the point where it does not have to depend on groundwater at all. With the Groundwater Act of B.E. 2520 and the government's stern efforts to halt groundwater withdrawal, the BMR's land subsidence and groundwater deterioration problems have been reduced.

The Public Health Act, B.E. 2484. This law was enacted in A.D. 1941. To prevent the spread of diseases and water contamination, it contains a section that prohibits discharging of waste and residue into public watercourses such as canals, rivers, and lakes. Violators are subject to a maximum 50-baht (US$2) fine, far too moderate to deter violators. This law is being revised to make it more effective.

The Industrial Work Act, B.E. 2512 (1969). This act is intended to govern all activities of industrial plants, including wastewater treatment and disposal of wastewater into the natural watercourse. An applicant for a permit to build a plant must submit an appropriate wastewater treatment plan, together with the application form. A review of the validity of this plan is an important element for approval. For those plants built without appropriate permits, the law sets a maximum fine of 100,000 baht (US$4,000) and cessation of all their operations. For those who fail to follow the approved water treatment plan, the plan sets a maximum penalty of 1 month in jail or 10,000 baht (US$400) fine, or both.

A majority of manufacturing and industrial plants are located in the BMR. Without this act and its effective enforcement, industrial effluent would be a major cause of the Chao Phraya River's pollution. The law has been amended several times and adjusted to changing situations, an indication that it is taken seriously. As stated earlier, the industrial sector is not considered a major source of the Chao Phraya River's pollution. The Industrial Work Act deserves much of the credit for this.

Effectiveness of Legal Means in Solving Water Use Conflicts

In practice, the laws previously described do not help much in solving or alleviating conflicts. There are many reasons for this.

First, ownership and water rights, although mentioned in the Civil and Commercial Code and the People's Irrigation Act, are not known by the general public. Hence, people rarely use them to resolve conflicts.

Second, many of the conflicts are not overt. For instance, people see a water shortage in the Central Plain's irrigation area as given by nature and under the government's management. They do not see it as a conflict between upstream and downstream uses.

Third, there has been, in actuality, no enforcement of some of the laws. For instance, the requirement that anyone who wants to irrigate more than 200 *rai* (16 ha) of land must obtain a permit from the authority, as specified in the People's Irrigation Act, is always ignored. The authority itself has not shown fortitude in implementing the law. Hence, water use conflict between the upstream Chiang Mai Valley and the downstream irrigation area in central Thailand is not taken care of by legal means.

Fourth, some laws are ineffective because of inadequate monitoring and reporting systems. Water pollution in the Chao Phraya River results largely from waste discharged from the household sector. Yet households in the BMR are so numerous that monitoring violations of each household is not possible.

Fifth, for some use conflicts, the lack of acceptable alternative sources of water prevents effective law enforcement. The case of excessive groundwater withdrawal and land subsidence is an example. The BMR was declared a groundwater management zone subject to regulation. According to the 1983 Cabinet Resolution, groundwater withdrawal within the BMR water supply service area, unless proved necessary, is prohibited since sufficient alternative sources of water are already provided for. Yet, the service area is only a small fraction of the BMR within the groundwater zone. For most of the BMR, leniency still prevails in permitting groundwater withdrawal.

Sixth, penalties in water-related laws are too lenient to be effective against existing problems. The 50-baht fine for discharging untreated effluents into a water body is too small to stop such violations, and it is not economical to establish an enforcement system to collect such a low penalty.

Seventh, it is natural that economic incentives can most of the time override the law. In the Chiang Mai Valley, economic development now comes first. People use water to meet the growing demands of development without trying to comply with the People's Irrigation Act.

Eighth, as more water upstream is withdrawn for development purposes, less is available for downstream uses, including irrigation. With less water, conflicts within the irrigation area covered by the Royal Irrigation Act, currently resolved by rotation of water delivery, will intensify.

Conclusion

Water use conflicts are evident in the BMR in various forms. A major cause is a rapid increase in demand for water, compared with a rather slow expansion of the supply side. Because of the spatial nature of water that flows from high to low land, water use conflicts in the BMR cannot be discussed without considering the adjacent and related areas.

Legal measures are a potential means for resolving or reducing conflict, since water is defined by law as belonging to the public. In this paper, all the laws relevant to the BMR water use conflicts were discussed. Legal means, although necessary, are not sufficient in resolving conflicts, however, because of economic pressures, low penalties, inadequate monitoring, and infrequent resort to the law to settle disputes. Therefore, other measures need to be explored. There is still some scope for traditional supply-side approaches, notably by building more storage dams and divisions and, in some places, by extracting more groundwater. But with a foreseeable limit on the supply side, demand management approaches will become more important. These include adopting water-saving technology such as drip irrigation for agricultural development. Furthermore, approximately 40 percent of the MWA water supply is unaccounted for, and a great deal of irrigation water is lost because of damaged canals. Hence, conflicts can be reduced by improving the distribution system to make more water available for other uses.

Where water becomes too scarce to grow rice, crops that require less water should be introduced. But this solution involves people's lifestyle, their households' food security, and markets for the introduced crops.

Annex:
Water-Related Laws in Thailand (B.E. 2500 = A.D. 1957)
(B.E. = *Buddhist era*; A.D. = *anno Domini*)

1. Thai Constitution, B.E. 2535
2. Civil and Commercial Code
3. Criminal Code
4. Royal Decree for Water Supply Establishment

5. Water Supply Canal Preservation Act, B.E. 2456
6. Municipal Act
7. Canal Preservation Act, B.E. 2444
8. People's Irrigation Act, B.E. 2482
9. Royal Irrigation Act, B.E. 2485
10. Ditch and Dike Act, B.E. 2505
11. National Energy Act, B.E. 2496
12. Electricity Generating Act, B.E. 2511
13. Municipal Electricity Act, B.E. 2501
14. Provincial Electricity Act, B.E. 2503
15. Northeastern Electricity Act, B.E. 2505
16. Thailand Waterway Navigation Act, B.E. 2456
17. Revolutionary Announcement on Ship Park in Canals, Rivers
18. Building Control Act, B.E. 2522
19. Ship Collision Prevention Act, B.E. 2522
20. Fishery Act, B.E. 2490
21. Fish Landing Act, B.E. 2496
22. Public Health Act, B.E. 2484
23. Industrial Work Act, B.E. 2512
24. Sanitary District Act, B.E. 2495
25. National Economic and Social Development Act, B.E. 2521
26. Confiscation of Fixed Asset Act, B.E. 2530
27. Thai Territory Fishing Rights Act, B.E. 2482
28. City Planning Act, B.E. 2518
29. Tidiness and Cleanliness Act, B.E. 2503
30. Housing and Land Control Act, B.E. 2504
31. Revolutionary Announcement on Highway Building
    and Maintenance
32. Municipal Water Work Act, B.E. 2510
33. Provincial Water Work Act, B.E. 2522
34. Groundwater Act, B.E. 2520
35. National Environment Conservation and Promotion Act, B.E. 2535
36. Agricultural Land Reform Act, B.E. 2518
37. Land Consolidation Act, B.E. 2517
38. Industrial Estate Act, B.E. 2522

References

Asian Institute of Technology. 1982. Ground Water Resources in Bangkok: Development and Management Study. National Environment Board, Thailand.

Camp, Dresser, and McKee. February 1987. Sewerage, Drainage and Flood Protection Systems, Bangkok and Thonburi, Thailand. Bangkok Municipality, Thailand.

Department of Industrial Works. 1987. Chao Phraya River Study 1986–1987, Division on Industrial Environment, Thailand.

NEB (National Environment Board). November 1983. Lower Chao Phraya Water Quality Study Project. Publ. 1982-003. Environment Quality Standard Division, NEB, Thailand (in Thai).

NESDB (National Economic and Social Development Board). June 1986. Bangkok Metropolitan Regional Development Proposals: Recommended Development Strategies and Investment Programs for the Sixth Plan (1987–1991). Joint NESDB/IBRD/USAID/ADAB Metropolitan Planning Project, Thailand.

Panswad, T., et al. November 1987. Domestic Wastewater and Water Pollution Problems in Bangkok and Its Vicinity. NEB Report No. 87/020, Thailand (in Thai).

Sharma, M. L. March 1986. Role of Groundwater in Urban Water Supplies of Bangkok, Thailand, and Jakarta, Indonesia. Environment and Policy Institute Working Paper, East-West Center, Honolulu, Hawaii.

TDRI (Thailand Development Research Institute). July 1988. Development of a Framework for Water Quality Management of Chao Phraya and Thachin Rivers. NEB 07-04-31, Office of the National Environment Board, Ministry of Science, Technology, and Energy, Thailand.

9

Water for Honolulu:
Use, Reallocation,
and Institutional Change

James E. T. Moncur

For Honolulu, the past century has brought profound growth in water consumption and sometimes alarming deterioration of water quality. Agriculture grew from subsistence farms to large, scientifically managed plantations and now is in decline under the challenge of urbanization. Change has been almost complete in terms of water sources and uses. The institutional framework underlying water allocation has undergone radical alteration in the past two or three decades by judicial, legislative, and administrative upheavals. As such, Honolulu[1] can serve as a microcosm of problems in water development, use, and transfer. The city thus provides an instructive case study of the effects of institutional change on economic efficiency of water use, as well as the effects of changes in water use on institutions.

Oahu's Water Economy

The City and County of Honolulu covers the entire island of Oahu and includes about 80 percent of the state's roughly one million resident population. Hawaii's economy depends heavily on tourism, federal government expenditures—mainly for military activities—and, to a declining extent, agriculture, principally sugarcane.

Figure 9.1 summarizes Oahu's water economy. The two surviving sugar plantations together account for nearly one-half of all water use, with other agriculture using an additional 2 percent. Sugarcane, a water-intensive crop, depends for viability on a structure of tariffs and import controls and is a declining industry. Nevertheless, there have been few significant water transfers from agricultural to urban uses.

153

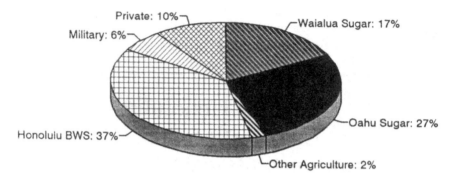

Total Consumption: 408 mgd

FIGURE 9.1 Oahu water consumption, 1988 (million gal/day) (*Source:* City and County of Honolulu 1990).

The 37 percent of total consumption delivered by the Honolulu Board of Water Supply (BWS) provides for a rapidly growing population, but few if any water-intensive industries or commerce other than perhaps hotels. Roughly 60 percent of BWS water goes to residential consumers; 25 percent to commercial, industrial, and hotel users; and the remainder to various government units (Honolulu Board of Water Supply 1982).

Groundwater provides nearly all the city's water. Most of the conventionally available sources have been or soon will be fully exploited. The state and city are cooperating to construct and operate a demonstration desalination plant that will produce water at three or more times the current water quantity charge. Some possibilities exist for water transfers from the windward to central parts of the island, but the quantities are limited, costs are high, and public opposition is intense. In short, the city is well into the rising portion of its long-run marginal cost curve.

Institutional Change

Traditional neoclassical theory considered the institutional context—the set of rules governing allocation and use of a commodity—as a parameter, specified outside the economic system and not subject to the actions of individual participants. Clearly, however, economic institutions emerge and evolve and sometimes die in response to changes in the demand-and-supply conditions for the commodities they allocate. They are not chiseled in stone (Olson 1965). This proposition is certainly true for water. As demand increases, water moves further and further beyond the status of a free good. Conflicts over use of a scarce resource arise, and the value of mechanisms to resolve those conflicts increases. One expects to see different types of water allocation arrangements where water is "scarce" than where it is "plentiful," because the value of water differs. Likewise, one expects to see water institutions evolve as demand grows relative to supply.

Markets are one institutional framework. Governmental direction of water use is in some sense the polar opposite of market-directed activity. Between these extremes lies a continuum of possibilities. Until recently, Hawaii's water allocation mechanisms could be seen as evolving gradually toward the market end of the spectrum, interrupted several times and in ways by regulatory or judicial decisions designed either to repair market failures or to wrest property rights in water use from one group to another.

The processes by which this evolution occurs have been much discussed in the literature on profit- and rent-seeking activity (Buchanan et al. 1980; Anderson 1983). Utility-maximizing individuals will compete to capture any potential rents they see by shifting around the resources they control. In some cases, this shifting occurs through political, rather than market, processes.

Development of Honolulu's Water Institutions

Traditional Hawaiians (say, prior to the Western "discovery" of the isles in 1778) appear to have simply shared the available supplies, with some oversight by the *alii* or noble classes. This allocation principle presumably sufficed for the purposes of the times, supporting a population thought to have been between 200,000 and 400,000 (State of Hawaii 1987:13) by the time Cook arrived. A relatively stable socioeconomic structure presented little impetus to change water allocations, although water was clearly valued highly in this mostly agrarian society.

The rise of sugar and expansion of taro cultivation in the mid-nineteenth century, however, greatly increased the demand for water and brought conflict that traditional institutions could not accommodate. The first step in resolving these conflicts was the "Great Mahele" of 1848, in which the kingdom's land was formally apportioned among the people. Along with the land went a presumptive right to use water. Sugar plantations, in particular, acquired water use rights and constructed systems of tunnels, canals, and ditches to move it to the most productive lands. Disputes over water were handled through court decisions. By 1920, water rights were well defined and clearly transferable between uses, users, and locations.[2]

Groundwater exploitation began with the drilling of a well by pioneer sugar planter James Campbell in 1879 (Cox 1981). This well proved highly productive, and many other water users followed Campbell's lead by drilling in the same aquifer. With no mechanism to control third-party effects, the predictable overdraft soon occurred. A declining water table and consequent saline intrusion forced abandonment of many coastal wells. Various commissions studied the problem, culminating in a 1929 recommendation to establish the Honolulu Board of Water Supply to operate the city's water system. That same year, in *City Mill vs. Honolulu Sewer and Water Commission*, the Hawaii Supreme Court pronounced the correlative doctrine for artesian water, more or less a groundwater analogue of riparianism in which overlying landowners are to share in using an aquifer.

The BWS was also given broad authority over water development for the entire island. It set standards, monitored well-drilling and operation, repaired leaking wells, and generally improved the efficiency of water extraction (Cox 1981:64–66). The resulting betterment of physical efficiency of the water system provided the island with supplies adequate to handle growing residential and military needs without major transfers from agricultural uses. In terms of competing demands for contemporary use, at least, raw water remained essentially a free good until recent years.

On some interpretations, by the time Hawaii gained statehood in 1959, the territorial courts had established firm and readily marketable water use rights (Anderson 1985). However, a suit filed on the island of Kauai in that year led to a decision that seemed to overturn completely one of the principal footings of market transferability. In the late 1940s, Gay & Robinson, a sugar plantation on the island of Kauai, replaced an old, leaky diversion channel with a new tunnel, and in the process reduced water available to McBryde Sugar Company. After a decade of negotiations, McBryde filed suit to clarify the water right. On appeal, the State Supreme Court ruled (*McBryde Sugar Co. vs. Robinson* 1958, 1973), to both parties' great surprise, that while both companies had appurtenant or riparian rights to use water, neither had a property right in it. Instead, the court declared the state to be the owner.

Gay & Robinson then sued the state in federal court (*Robinson vs. Ariyoshi* 1982), claiming its property had been taken without just compensation in violation of the U.S. Constitution. Action in this case proceeded through federal courts, where the most recent ruling, in September 1989, favored the state on technical grounds but did not settle the substantive question of whether a taking had occurred.[3] However, the matter is still not entirely settled. The court declared that although the landowners' title to water had indeed been "clouded" by the *McBryde* ruling and subsequent state action, no physical taking had occurred and the issue was deemed to be not yet "ripe for review."[4] Presumably, some future state action in approving or denying a water permit, or the transfer of a permit, will make the case "ripe" and generate a definitive ruling.

Based in part on the *McBryde* decision, the court subsequently determined in *Reppun vs. Board of Water Supply* that while the owner of a water right might give up that right, no one could receive it, hence essentially quashing the notion of transferability (Chang and Moncur 1984).

The 1987 Hawaii State Water Code

The *McBryde* decision and ensuing controversy led the 1978 Hawaii Constitutional Convention to propose an amendment,[5] approved by voters later that year, directing the state legislature to establish an agency to set water conservation, quality, and use policy to protect the state's water resources and to regulate water use. There followed almost a decade of studies, proposals, conferences, hearings, and discussions. Major proposals from official study commissions would have required all users to obtain permits of limited duration from a state authority with powers to deny renewal and to disapprove any transfers (Governor's Advisory Commission 1979). The Legislative Reference Bureau Advisory Commission went so far as to denounce market institutions: "Indeed, the commission strongly feels that water and the right to use it should not be the subject of purchase and sale on the open market" (Legislative Reference Bureau 1986). This sentiment carried over into early drafts of the legislative act creating the code, though it was not a part of the final act.

The resulting State Water Code (Hawaii Rev. Stat. §174C, Supp. 1989) has been called a masterpiece of compromise (Lau 1988). What remains unclear is whether the compromise melded the best or the worst of competing paradigms. I will try to summarize provisions of the code and then to evaluate its economic efficiency and equity effects.

Provisions of the Code

The new law establishes a State Water Resources Management Commission with planning and managerial functions. The commission designates a Water Management Area wherever water resources are threatened in quantity or quality. Water outside these areas is not subject to regulation, although all water users must file a declaration of their uses.

The allocation system established by the code centers on a system of water use permits. Within the designated areas, any water use except domestic applications requires a permit, with existing uses grandfathered. Once issued, permits are of essentially unlimited duration, specifying the water source, quantity, use, location, and other information requested by the commission. The commission may allow the permittee to transport

the water outside the overlying land or watershed, but must approve almost all modifications of permit terms. County water utilities, however,
need approval only for changes in quantity or changes involving third-
party effects.

A major and difficult responsibility of the commission is protecting
water quality and instream uses. Thus far, the rule generally has been
simply to allow no degradation. Also, the code specifies protection of Native Hawaiian water rights, a problem as difficult and controversial for
Hawaii as some Native American water rights issues have been on the
U.S. mainland.

Evaluation of the Water Code

Howe, Schurmeier, and Shaw (1986) developed six criteria for a system of allocating water use rights. These criteria grew from standard
principles of economic efficiency and may be summarized as follows:

- Is the system *flexible* in accommodating changes in climate, demo-
graphic factors, and economic conditions?
- Does the system ensure *security of tenure* of the right?
- Do rights holders face the *full real costs* of using the water?
- Are the outcomes of the allocation process *predictable*?
- Is the allocation process perceived as *equitable* or fair?
- Does the system reflect *public values* not adequately considered by
individual water users?

Space and time do not permit a full discussion of all these criteria (see
Moncur 1989b). In terms of economic efficiency, however, flexibility probably is the code's most important weakness. The code allows transfers of
water permits if the place, quantity, and purpose of the use, and possibly
other conditions as well, remain unchanged. The commission may require other changes in the permit as a condition for approving a requested modification. Any transfers involving a change in use, time, or
location—and these are the principal sources of potential efficiency
gains—must undergo the full application process. In principle, the commission must consider competing applications along with any others involving existing permit holders attempting to transfer their use rights. By
raising transaction costs, all these factors reduce the likelihood that permit holders will apply for a transfer or modification.

In early proposals for the code, it was asserted that the flexibility to be
maximized was that of the state in reallocating the water (Legislative
Reference Bureau 1986:19). To this end, the permits were to be issued for
as short as 5 years. A draft of the water code sent to the House-Senate
Conference Committee in February 1987 included a clause explicitly pro-

hibiting the sale of water covered by permits. The committee deleted this provision, but warned in its report that its action "does not imply that the 'sale' of water is affirmatively sanctioned" (Hawaii Legislature 1987). Thus the commission presumably has some latitude in its treatment of market-inspired transactions. The commissioners' interpretation of this point will be as important to efficiency of water allocation as is the formal code itself.

To the degree that sale or rental of water use rights is prohibited or successfully discouraged, permit holders will tend to persevere in low-valued uses as long as they can reap positive returns to the water, however much that return may fall below some alternative use. Likewise, users have an incentive to use more water than necessary just to preserve their rights. Indeed, under a similar authority granted by the state's Groundwater Control Act of 1979, which was largely incorporated into the new code, the state in 1985 cited nonuse and decreased the water allocation of Oahu Sugar Company by 20 million gallons per day, in favor of the county Board of Water Supply and a military installation. The plantation argued, unsuccessfully, that it had not used the groundwater only because of unusually high flows from its less costly surface water sources during the preceding 2-year period (Balfour 1988:2).

In a recent assessment of sustainable yield in one of the island's principal aquifers (Yuen, Mink, and Chang 1988), it was recommended that the previous estimate of 225 million gal/day be reduced to 195. Although the latter figure covers current uses, all 225 million gal/day have been allocated. Commission staff recommendations, approved in March 1989, called for decreasing allocations to 195 million gal/day by 1995, with most of the decrease (84 percent) coming from Oahu Sugar Company. No provision was made for compensation, so the commission appears to have denied that Oahu Sugar had any property rights in the water. This controversy will be compounded by massive housing and resort developments projected for the area served by this aquifer.

One cannot expect perfection in real-world institutions. The question is whether the new water code improves or hinders attainment of the efficiency criteria. The code is still a relatively new and incompletely formed institution. The law does not explicitly define a water right and gives the Commission on Water Resources Management wide latitude on many points of its mandate. As of late 1990, the commission has organized itself, adopted detailed procedural rules and regulations, approved interim instream flow standards, and approved many permit applications covering precode established uses. None of its actions has generated a court challenge. Some sort of de facto water use right (and conditions of transfer) will emerge over time as the commission compiles a record of decisions and the courts interpret those decisions. Economic efficiency of the resulting system may differ significantly from what appears on the face of the law.

Rent-Seeking Activity

Tullock defines *rent seeking* as "the use of resources in actually lowering total product though benefiting some minority" (Tullock 1989:vii and Chapter 6) where "product" should be understood to include nonmarket as well as market benefits. Examples include funds spent lobbying for a tariff or for a grant of monopoly power or exclusive access to some resource. Much rent seeking is oriented toward the exercise of government power.

Prior to 1973, water rights in Hawaii were readily transferable and frequently traded. Rents attributable to these rights were no doubt positive in many cases. Since the transfers were voluntary, they would have been Pareto improving, not rent seeking. Water was thus moved from low value to higher value uses as buyers and sellers saw mutual benefit in transactions. By contrast, vesting allocational authority in government can be expected to induce rent-seeking activity, with a lowering of total product as described in Tullock's definition. The case for inefficiency and litigiousness of government allocation processes is well-documented (e.g., see Trelease 1974 or Anderson 1983). The BWS, housing developers, and small farmers, for example, can be expected to continue to seek favored status (in forms such as obtaining new water rights at less than full economic cost, or retaining below-cost water rates), whatever the economic efficiency of their current or planned uses.

Whatever the State Supreme Court's motives and reasoning, the assertion of state ownership from *McBryde* was entirely unexpected. An editorial writer commenting on a related later ruling noted that "The [Hawaii Supreme Court] flabbergasted everybody involved in a fight already 50 years old by declaring the water didn't belong to either plantation; it was public property. No one expected this outcome to what seemed to be a purely private fight" (Smyser 1989:A14). One may think of *McBryde*, then, as a random shock to the existing system of water rights. Although it applied directly only to surface waters, this decision threw into question the commonly held assumptions regarding ownership and control of water resources. Long-standing water users saw their rights muddied; for others, the decision opened the possibility of access to or control over a resource of great and growing value. *McBryde*, then, was an invitation to rent-seeking behavior.

One example of such rent seeking is provided by *Reppun vs. Board of Water Supply* (1982). The BWS had purchased water rights with the intention of diverting water to another basin. In the wake of the *McBryde* decision, Reppun, representing a group of small taro farmers, sued to halt the diversion, arguing that water flows coming through their taro patches would be substantially diminished. In deciding in favor of the farmers, the court did not even consider economic efficiency, though the diversion clearly met standard efficiency criteria (Chang and Moncur 1984).

Similar influences were brought to bear in the 1978 Constitutional Convention, where one group wished to follow the lead of California in declaring that all waters in the state were owned by the state.[6] In the final compromise, a constitutional amendment was proposed, as noted earlier, mandating the legislature to provide for the regulation of water.

The amendment spawned further possibilities for rent seeking. Activity and debate leading up to passage of the code in 1987 have already been noted. The main proponents of allocation by government, or variations thereof, included speakers for small farmers and the union representing most sugar plantation workers. State agencies took no formal position, but some individuals in the Department of Land and Natural Resources, the agency that eventually was given principal regulatory authority, argued in favor of limited duration permits. County water supply agencies opposed proposals vesting allocation authority in the state, but similarly opposed entrusting water allocation to markets. Advocates of more market-oriented approaches came principally from landowner groups—the sugar planters who had, before 1973, controlled much of the state's developed water.

As noted in Figure 9.1, nearly one-half of total water use on Oahu goes to the two surviving sugar plantations. Continued use of water for sugar is uncertain for reasons largely unrelated to water. Plantations almost surely will cease growing cane if federal sugar import quotas are abolished or reduced significantly. Urbanization threatens the existence of Oahu Sugar Company, whose cultivated lands are all leased. Cessation of sugar might release a great deal of water for alternative uses. Several minor crops might replace sugar on some of these lands—coffee, macadamia nuts, and pineapple are often mentioned—but these would probably use less land than sugar and would surely use less water per unit of land. The demise of other sugar plantations on Oahu and other islands has left much of the former sugar land unused. Golf courses are a possibility for some of this land. Golf courses, which use less water than sugarcane (City and County 1990:155), can use brackish water, where available, freeing current potable irrigation water for domestic purposes.

However, under *McBryde* and the State Water Code, the ability of the plantations to profit from the sale of water rights they now control is uncertain at best, and transaction costs have risen substantially. Hence the plantations have only limited incentives to pursue efficiency-enhancing water transfers. Assuming they cannot sell the water rights, profit maximization would call for using water up to the point where its marginal product falls to zero or other factors force them out of sugar.

Water Pricing

Urban water agencies in the United States have tended to considerably underprice water,[7] thus in effect encouraging wasteful use of an increasingly scarce resource and signaling premature development of high-cost water sources such as desalination or distant interbasin transfers. Prices, which are usually set based on average cost, are understated for several reasons: (1) failure to adjust the cost base for inflation; (2) failure to include equity capital and the value of donated water system components in the cost base; and (3) ignoring the scarcity value of raw water in situ.[8] In the case of Honolulu, incorporating these elements into the rate base would justify raising the quantity charge by a factor of at least two.

Of course, the conservation advantages of correct pricing depend on users bearing this price. Although single family housing in Honolulu is fully metered, most apartments and commercial customers have only a master meter for the entire building or complex, with water charges passed on as an unspecified element of rent or maintenance fees. Whether the elasticity of demand for water by apartment dwellers is high enough to induce significant decreases in water use, and thus justify the costs of metering, remains an open question. Other pricing mechanisms, however, may allow true economic costs to be passed along to users in such a way as to induce conservation. Development charges, in particular, could be assessed in such magnitude as to encourage greater housing density, less yard space, and thus lower water consumption.

Also, current Honolulu rates are uniform islandwide, while costs vary widely, especially with altitude in this very hilly city. Since hilltop residents are generally better off than average, this "postage stamp" pricing scheme results in the perverse subsidization of wealthy customers.

Finally, the Honolulu water system has faced two periods of drought in the last 15 years. In both cases, the BWS imposed a program of use restrictions, but left water rates unchanged. The State Water Code (Hawaii Rev. Stat. §174C, Supp. 1989) empowers its Water Commission to declare a shortage and to "apportion, rotate, limit or prohibit" water use in an area. Evidence suggests that some form of drought surcharge would bring about the desired conservation, without some of the negative consequences of restrictions (Moncur 1987, 1989a).

For all the modifications in water pricing that seem clearly justified, however, the real question may lie in the incentives provided by the water supply institution for managers to set price at the economically efficient level. As a government agency, managers and their board of directors face no economic pressure except maintaining a cash flow adequate to the system's "needs." On the other hand, they face considerable political pressure from user groups and city politicians to keep water rates as low as possible. For example, the Honolulu BWS has had a preferential rate for the 1 or 2 percent of its output going to small agricultural users. Two major rate studies in the past 15 years have recommended rescinding this preference. Resistance by small farmers, however, has resulted in rate increases smaller, in relative terms, than those applied to other users. At the same time, managers—typically with engineering or technical backgrounds—are by training predisposed to supply augmentation solutions rather than demand management. The value of engineers' vested human capital relies largely on continuing to build structural solutions to water problems.

Water Quality

As with many other cities, Honolulu has gone through several episodes of intense concern with water quality in the past two decades. Controversy has centered on polluted water sources and wastewater treatment. Groundwater provides almost all of the city's water sources, and with careful management of watershed lands there has been no need for chlorination or other purification. Recent improvements in the detection of pollutants, however, have turned up evidence of minute levels of certain chemicals in some drinking-water sources in central Oahu. Although it seems likely that the pollutants came from agricultural sources, such important details as travel times and specific places of origin of the chemicals have not been determined. Effects of these chemicals are not well established nor are threshold levels for generating human harm. Nevertheless, an activated carbon filtration plant was built and is operating to remove them, and substantial investments were made in quality-monitoring facilities. Costs for this plant were built into the general BWS budget and passed along to customers as part of the islandwide uniform water rate. Neither the direct beneficiaries nor the pollutant generators have paid any of these specific costs themselves.

As for wastewater, the U.S. federal government has recently mandated secondary treatment of sewage, although waivers can be obtained where a locality can show that primary treatment suffices. For Honolulu, treatment plant effluent is dispersed in deep ocean outfalls at some distance from shore, and secondary treatment would provide little added benefit (J. Harrison, Environmental Coordinator, University of Hawaii Water Resources Research Center, pers. com.). Hence the city applied for a waiver to avoid estimated costs of roughly US$500 million for secondary treatment at its three major sewage plants.

However, the sewage collection systems are not entirely impervious to runoff from especially heavy storms, and during such times some of the sewage lines can overflow. In several instances, sewer lines spilled into Kaelepulu Stream, which courses through residential areas and empties into the ocean through the middle of a popular beach park. The mouth of the stream is shallow and gentle and has become a popular spot for children. The prospect of overflow going into this environment so aroused residents of the suburb through which Kaelepulu flows that the city agreed to withdraw its application for waiver of secondary treatment. The cost will amount to well over US$100 million.

On the benefits side, little or nothing is known about consumers' willingness to pay for added water purity of the sort secondary treatment will provide. However, in the case just described, the costs will be evenly distributed throughout the entire city as a portion of the total sewage treatment bill. Beneficiaries—residents near the treatment plant and ardent environmentalists—will clearly avoid paying most of it. The federal government may fund a substantial portion of the costs, thus further diluting and disguising these costs from all taxpayers, beneficiaries, and others alike. Even though benefits are only a fraction of costs, they accrue mainly to a relatively small group of people who feel it worth their while to use the political process to obtain the benefits. Costs, by contrast, come to rest only in a small degree on any one person or group, so that no one has much incentive to oppose the project. As with the water filtration example, there is no institutional mechanism to ensure that costs come to bear on the same persons or groups who prompt construction of the facilities and reap most, if not all, of the benefits. There is, to borrow a term from dentistry, a malocclusion between benefits and costs, a mismatch inviting rent-seeking activity.

Conclusions

If there has been a theme to all this, it has been that the economic efficiency of water use in Hawaii emerges from political activity rather than open, competitive market operations, much more so than for most other goods and services and more so in Hawaii than in some other U.S. jurisdictions. Incentives to conserve are diminished or absent due to mispricing, which in turn arises from absent or misplaced incentives to managers; infrastructure for both quantity and quality is probably overbuilt due to mismatched incidence of costs and benefits. The new water code probably fails to remedy any of these problems.

It is not that Honolulu will come to the end of its water supplies at some future day and thereafter have to turn people away at the airport. But current institutions seem bound to make water and its treatment much more expensive than necessary, thereby diverting expenditures from other desirable goods and services, whether public or private. And, more subtly, this illusory water "scarcity" may alter the path of general economic development.

Of course, conditions in other cities, Asian or American, differ from Honolulu. However, "political economy" problems of water differ in degree more than substance. As a broad generalization, notions of water marketing and the efficiency to be had thereby are gaining in most parts of the United States.[9] One cannot feel certain about the directions in which Hawaii has moved recently, but it would seem that Hawaii provides a series of avenues not to be emulated.

Notes

1. Hawaii is unique among the U.S. states in having only two levels of government, state and county. The City and County of Honolulu governs the island of Oahu, and its semi-independent Board of Water Supply is the sole purveyor of urban water.

2. T.L. Anderson in "The Market Alternative for Hawaiian Water," *Natural Resources Journal*, Vol. 25, No. 3, October 1985, pp. 893–909, cogently argues that market processes served Hawaii well, at least until judicial intervention in the 1970s.

3. "It seems clear that the issues commented upon by the Hawaii Supreme Court [in the 1973 *McBryde* ruling] substantially clouded the title of the appellees and could affect financing and transfers of property interests. . . . However, we cannot say at this point that appellees 'retained [no] beneficial use' or that their 'expectation interests had been [completely] destroyed. . . .' We therefore conclude that even if the State of Hawaii has placed a cloud on the title of the various private owners, this inchoate and speculative cloud is insufficient to make this controversy ripe for review." *Robinson vs. Ariyoshi*, 887 F.2d 215 (9th Cir. 1989).

4. *Ibid.*

5. Hawaii State Constitution, Article XI, Section 7:

The State has an obligation to protect, control and regulate the use of Hawaii's water resources for the benefit of its people.

The legislature shall provide for a water resources agency which, as provided by law, shall set overall water conservation, quality and use policies; define beneficial and reasonable uses; protect ground and surface water resources, watersheds and natural stream environments, establish existing correlative and riparian uses and establish procedures for regulating all uses of Hawaii's water resources.

6. This and the following paragraphs rely on an interview with Richard H. Cox who, during the period leading up to passage of the Code, was vice-president of a sugar-growing and landowning company and who currently is a member of the State Water Resources Management Commission.

7. To achieve economically efficient rates of resource use, water (or any other commodity) should be priced at marginal cost, where marginal cost includes extraction, distribution, and administrative costs as well as scarcity rents. The "efficiency price" of a resource is given by D.A. Hanson, "Increasing Extraction Costs and Resource Prices: Some Further Results," *Bell Journal of Economics*, Vol. 11, No. 1, Spring 1980, pp. 335–341. On marginal cost pricing in general, see J. Hirshleifer, J.C. DeHaven, and J.W. Milliman, *Water Supply: Economics, Technology and Policy* (Chicago: University of Chicago Press, 1969). Urban water rates in the United States seem also to be considerably below those of most other developed nations (see *The Economist*, December 1, 1990, p. 119).

8. The first two points are discussed in J.E.T. Moncur and R. Pollock, "Accounting induced distortions in public enterprise pricing," paper presented to Western Economic Association, Los Angeles, CA, June 1988. The third is the subject of J.E.T. Moncur and R. Pollock, "Scarcity Rents for Water: A Valuation and Pricing Model," *Land Economics*, Vol. 64, No. 1, February 1988, pp. 62–72.

9. B.C. Saliba and D.B. Bush in *Water Markets in Theory and Practice: Market Transfers, Water Values, and Public Policy* (Boulder: Westview Press, 1988), document and evaluate a growing trend toward the use of markets or market-like institutions for transferring water rights in the Western United States.

References

Anderson, T. L., ed. 1983. *Water Rights: Scarce Resource Allocation, Bureaucracy, and the Environment*. Cambridge, Mass.: Ballinger Publishing.

Balfour, W. D., Jr. March 22, 1988. Testimony to State Water Commission on Proposed Administrative Rules, Honolulu, Hawaii.

Buchanan, J. M., R. D. Tollison, and Gorden Tullock, eds. 1980. *Toward a Theory of the Rent-Seeking Society*. College Station, Tex.: Texas A&M University Press.

Chang, W. B. C., and J. E. T. Moncur. September 1984. *Reppun v. Board of Water Supply*: Property Rights, Economic Efficiency and Ensuring Minimum Streamflow Standards. Technical Report No. 165, University of Hawaii Water Resources Research Center, Honolulu.

City and County of Honolulu. March 1990. Oahu Water Management Plan. Department of General Planning, Honolulu, Hawaii.

City Mill v. Honolulu Sewer and Water Commission, 30 Hawaii 912. 1929. Honolulu, Hawaii.

Cox, D. 1981. "A Century of Water in Hawaii," in F. N. Fujimura and W. B. C. Chang, eds., *Groundwater in Hawaii: A Century of Progress*. Honolulu: University Press of Hawaii.

Governor's Advisory Commission. January 1979. Hawaii's Water Resources: Directions for the Future. Honolulu, Hawaii.

Hawaii Legislature. April 27, 1987. Conference Committee Report No. 118 on H.B. No. 35, H.D. 1, S.D. 2, "A Bill for an Act Relating to the State Water Code." State of Hawaii.

Honolulu Board of Water Supply. July 1982. Oahu Water Plan, 4th ed. Honolulu, Hawaii.

Howe, C. W., D. R. Schurmeier, and W. D. Shaw, Jr. 1986. "Innovative Approaches to Water Allocation: The Potential for Water Markets." *Water Resources Research* 22(4): 439–445.

Lau, L. S. 1988. "State Water Code—A Masterpiece of Compromise." *Wiliki O Hawaii* 23(8): 1–3; 23(9): 2.

Legislative Reference Bureau. January 1986. Advisory Commission on Water Resources. Thirteenth State Legislature, Honolulu, Hawaii.

McBryde Sugar Co. v. Robinson, 54 Hawaii 174, 1958; 504 P.2d 1330 aff'd on rehearing, 55 Haw. 260, 517 P.2d 26, 1973.

Moncur, J. E. T. 1987. "Urban Water Pricing and Drought Management." *Water Resources Research* 23(3): 393–398.

———. 1989a. "Drought Episodes Management: The Role of Price." *Water Resources Bulletin* 25(3): 499–505 (June).

———. 1989b. "Economic Efficiency and Institutional Change in Water Allocation: The 1987 Hawaii Water Code," in F. E. Davis, ed., *Water: Laws and Management*. Tampa, Fla.: American Water Resources Association. A compilation of papers presented at the 25th Annual Conference, AWRA, 17–22 September, Tampa, Florida.

Olson, Mancur. 1965. *The Logic of Collective Action*. Cambridge, Mass.: Harvard University Press.

Reppun v. Board of Water Supply, 65 Hawaii 531. 1982. Honolulu, Hawaii.

Robinson v. Ariyoshi, 65 Hawaii 672. 1982. Honolulu, Hawaii.

Smyser, A. A. 17 September 1989. Richardson Court Bent Rules in Public's Favor. *Honolulu Star-Bulletin*. P. A14. Honolulu, Hawaii.

State of Hawaii Data Book. 1987. Honolulu, Hawaii.

Trelease, F. J. 1974. "The Model Water Code, the Wise Administrator and the Goddam Bureaucraft." *Natural Resources Journal* 14:207–229.

Tullock, G. 1989. *The Economics of Special Privilege and Rent Seeking*. Boston: Kluwer Academic Publishers.

Yuen, G., J. Mink, and J. Chang. May 1988. Review and Re-evaluation of Groundwater Conditions in the Pearl Harbor Groundwater Control Area. Report R-78 prepared for the Hawaii Department of Water and Land Development, Honolulu, Hawaii.

10

Water Use Conflicts Under Increasing Water Scarcity: The Yahagi River Basin, Central Japan

Kenji Oya and Seiji Aoyama

Japan is blessed with rich water resources, with an average annual precipitation of about 1,800 mm. In per capita terms, however, its per capita precipitation is only about one-sixth of the global average of 34,000 m³/year. Also, because of their short length and high runoff rates owing to the steep topography, rivers in Japan are not inherently conducive to large-scale water resources development. Nonetheless, surface flows meet about 80 percent of Japan's aggregate water demand.[1]

Urban domestic and industrial water demands in Japan grew at a phenomenal rate from rapid economic growth during the postwar period (Table 10.1). As the base flow of rivers had already been almost entirely claimed for agricultural uses, water supplies for domestic and industrial uses were enhanced through the construction of storage reservoirs, especially in major metropolitan regions such as Tokyo, Osaka, and Nagoya. A number of pieces of legislation were enacted in the 1950s and 1960s to facilitate river basin development while respecting the vested rights of irrigated agriculture.

Domestic water demands in the metropolises have continued to increase because of changes in lifestyle, coupled with rapid proliferation of the tertiary sector. In contrast, the demand for industrial water has been declining since the early 1970s, when the water-using industries began to adopt less water-consuming production processes and more in-plant water-recycling systems. This change was primarily due to the increase in water prices that followed the first oil crisis in 1973. Moreover, the stringent control measures enforced over water pollution and groundwater abstraction have also contributed to the increased recycling of industrial

TABLE 10.1   Changes in Water Use by Sector, Japan (100 million m³/yr)

Sector	1965	1970	1975	1983	2000[a]
Domestic water	42	69	96	121	208
Agricultural water	—	—	570	585	626
Industrial water	122	173	173	158	222
Total	n.a.	n.a.	839	864	1,056
Gross industrial water use	189	n.a.	486	554	893
Rate of recycling (%)	35.5		64.5	73.3	76.7

[a]The figures for year 2000 are projected values.

Source: National Land Agency 1987.
n.a. = Not available.

water (Table 10.1). The rate of water recycling in the industrial sector has been about 70 percent in recent years.

This paper attempts to review, through a case study approach, the responses to water use conflicts that have emerged as a consequence of increased water use in Japan's metropolises. More specifically, it briefly investigates the main features of the water use systems developed in the Yahagi River basin, and proceeds to analyze the problems of (1) water allocation during periods of drought, and (2) water quality deterioration and its influence on the choice of sources for irrigation water.

Water Use in the Yahagi River Basin: An Overview

The Yahagi River basin has an area of 2,260 km² and comprises some twenty-six municipalities (cities, towns, and villages) in the three prefectures of Aichi, Gifu, and Nagano in the eastern part of the Nagoya Metropolitan Region. While most of these municipalities are in Aichi Prefecture, part of the upper reaches of the Yahagi are in Gifu and Nagano prefectures (Figure 10.1). The upstream area, which accounts for about 70 percent of the total surface area of the river basin, is hilly and mountainous with scattered village communities. The midstream contains one of the leading agricultural areas in Japan, with extensive irrigation networks. It is also the site of heavy industrial concentration, with the giant Toyota automobile headquarters as the core. The downstream plain areas of the river basin have been undergoing rapid urbanization due to the expansion of the Nagoya Metropolitan Region in the west (Table 10.2).

The total water consumption per year in the Yahagi River basin is estimated at 900 million m³. Of this, about 100 million m³ is groundwater

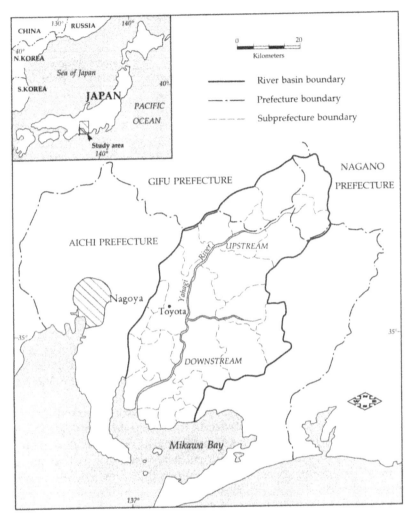

FIGURE 10.1 Yahagi River basin.

and some 150 million m³ is recycled from the midstream and downstream irrigation systems. Hence, the net volume of water supplies
directly dependent on the Yahagi River is estimated at about 650 million m³.

In a dry year, the total runoff from the entire river basin is estimated at
1,900 million m³, with 1,300 million m³ coming from the mountainous
catchment area. Thus, during dry years, the rate of river water utilization
is as high as 50 percent of the catchment inflow.[2] In 1978, when a serious
drought occurred in the Yahagi River basin, a survey recorded that over

TABLE 10.2   Changes in Population and Employment by Sector in the Yahagi River Basin

Year	Sector	Population ('000)	Employment Structure			
			Primary	Secondary	Tertiary	Total
1955	Yahagi River Basin	709	144	103	82	329
	Aichi Prefecture	3,769	525	470	476	1,471
	National total	89,276	16,111	9,220	13,928	39,259
1980	Yahagi River Basin	1,179	55	288	202	545
	Aichi Prefecture	6,222	166	1,292	1,589	3,047
	National total	117,060	5,770	19,260	30,330	55,360
1955–80	Yahagi River Basin[a]	166	38	280	246	166
	Aichi Prefecture[a]	165	32	275	334	207
	National total[b]	131	36	209	218	141

[a]1955 as 100 percent
[b]1970 as 100 percent

Source: Aichi Prefecture Government, various years.

60 percent of river water was utilized for irrigation during the peak water-consuming months (May to September). Thus, the Yahagi can be regarded as a highly exploited river. Only a few other rivers in Japan have utilization rates exceeding 40 percent. These include the Tone River in the Capital Region, as well as some rivers in the Seto Inland Sea Region and the northern part of the Kyushu Island.

Water use in the Yahagi River basin has increased markedly since the early 1970s. The largest contributor to the increased water demand is the agricultural sector, followed by the domestic sector (Table 10.3). The main reason for the increased agricultural water use is that the time-honored practice of repeated use of return flows for irrigation declined considerably due to water quality deterioration. In contrast, demands for industrial water fell during the 1970s, despite the rapid development of automobile and related industries, due to the widespread adoption of water-recycling technologies.

According to the 6th Aichi Prefecture Regional Plan adopted in March 1989 (see Aichi Prefecture Regional Planning Committee 1989:264–269), the aggregate water demand in the Yahagi River basin is projected at about 1,280 million m^3 in the year 2000, an increase of some 200 million m^3 over the current water supply capacity. Water use in the river basin has stabilized in recent years due to the slower growth of the regional economy, together with the widespread adoption of water-saving technologies. In the long run, however, water demand, particularly in the domestic sector, is anticipated to grow at a faster rate because of the improved levels of living and a growing population.

TABLE 10.3   Water Demand by Sector and Source, Yahagi River Basin (million m³/yr)

Water Demand	1970[a]		1978[a]		2000[b]	
		(%)		(%)		(%)
Total demand	841.5	(100.0)	1,018.1	(100.0)	1,276.1	(100.0)
Sector						
Domestic	91.7	(10.9)	127.3	(12.5)	236.9	(18.6)
Industrial	176.7	(21.0)	145.6	(14.3)	197.5	(15.5)
Agricultural	526.8	(62.6)	671.9	(66.0)	769.4	(60.3)
Fish farming	46.3	(5.5)	73.3	(7.2)	72.3	(5.6)
Source						
Main stream	286.1	(34.0)	496.8	(48.8)	722.7	(55.4)
Tributaries						
and ponds	377.8	(44.9)	381.8	(37.5)	500.2	(38.4)
Groundwater	177.6	(21.1)	139.5	(13.7)	80.2	(6.2)

[a]Aichi Prefecture Government, various years.
[b]Data drawn from Aichi Prefecture Regional Planning Committee 1989.

Two approaches have been adopted to cope with the projected future water demands. One is to further augment the water supply capacity by constructing an additional storage dam in the upper catchment area and a barrage at the mouth of the river. The second approach is geared to improving the efficiency of water use by (1) educating water users in water-saving practices, (2) linking together reservoirs, and (3) promoting water recycling and reuse of sewage.

Development of Water Use Systems

A brief review of the history of water resources development in the Yahagi River basin is in order for a better appreciation of water use problems the basin is facing today. Two distinctive stages can be identified. The first, marked by the exhaustive use of the base flow for irrigation, took place in the latter half of the nineteenth century. The second stage has been characterized by the postwar policy efforts to augment water supplies through development of reservoirs (Figure 10.2).

Development of Extensive Irrigation Systems

Generally speaking, agricultural development in the river plains of Japan increased remarkably between the sixteenth and the eighteenth centuries, as technologies became available for large-scale river works and flood control. Agricultural development in the Yahagi plain occurred somewhat later. In the late nineteenth century after the Meiji Restoration

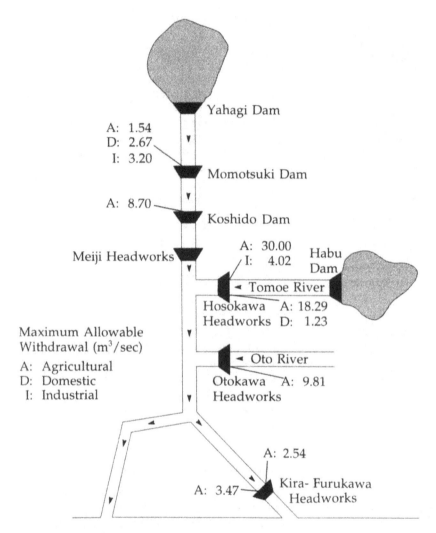

FIGURE 10.2 Water use systems in the Yahagi River basin (*Source:* Aichi Prefecture Government 1989).

(1868), two large-scale irrigation systems were developed on the flood plains along the midstream of the Yahagi. They are known as the Meiji and Shidare irrigation systems, covering up to about 8,000 and 2,000 ha with maximum allowable withdrawals of 30 and 8.7 m³/sec, respectively (see Table 10.4).

With irrigation, the plains areas were transformed rapidly from a barren plateau into a center for rice production with advanced mixed farm-

TABLE 10.4 Irrigated Areas in the Yahagi River Basin

Irrigation Area	Rice Field	Upland Field	Total	Maximum Allowable Withdrawal (m³/sec)
Shidare Yosui	2,000	—	2,000	8.69
Yahagi North	500	310	810	1.54
Meiji Yosui (right bank)	6,810	30	6,840	30.00
Nanbu (Meiji left bank)	850	350	1,200	1.44
Yahagi Downstream	7,327	417	7,744	16.85
Total	17,487	1,107	18,594	58.52

Source: Aichi Prefecture Government 1989.

ing, and eventually became a progressive agricultural area known throughout the country as the "Denmark of Japan." In the early 1930s, while most Japanese farming communities were swept into impoverishment by the world depression, the Yahagi plains remained the center of a great deal of activity. Many people interested in rural reconstruction and development visited the area (see Meiji Irrigation System 1979:211–232).

In terms of water use, the irrigation systems developed in the midstream of Yahagi were so extensive that almost the entire base flow of the main course of Yahagi was taken for agricultural use. For example, the maximum allowable withdrawal of the Meiji irrigation system (30 m³/sec) is almost equivalent to the base flow measured at its headworks. As in other river basins of Japan, the right to withdraw such a large volume of water was granted to the irrigation associations under the River Act of 1896 as a customary right. This right was reaffirmed by the provisions of the new River Act of 1964.

Augmentation of Water Supplies Through Reservoir Development

The water rights granted to the two irrigation systems had a decisive influence on the development of water use systems in the postwar period. As the base flow of the Yahagi River had already been fully claimed, additional water supplies had to be provided by developing reservoirs. Much of this increment also went to agriculture.

To date, two major reservoirs have been built in the Yahagi River system. One is the Habu reservoir built in 1963 on the Tomoe River, a tributary of the Yahagi, as an exclusive source of irrigation water. Another is the Yahagi multipurpose reservoir, constructed in 1971 on the upper reaches of the main course of the Yahagi. This reservoir is designed to augment water supplies for agriculture, industry, and domestic uses, and to generate hydroelectric power and control floods.

The Habu reservoir supplies water to irrigate an area of about 7,700 ha in the coastal alluvial plain downstream of the Meiji and Shidare irrigation systems. This area, originally a swampy delta at the mouth of Yahagi River, was brought under rice cultivation in the seventeenth and eighteenth centuries. Using small irrigation facilities drawing from the mainstream of the Yahagi River and its tributaries, farmers were not as much concerned with the availability of water as with flood control and drainage. At the time, the area was lower than the river bed. Accordingly, when construction of the Meiji and Shidare irrigation systems was initiated, the downstream communities did not file claims for a share of water in the main course of the Yahagi.

With upstream development, primarily the construction of hydroelectric dams upstream of Yahagi, the downstream farmers experienced severe water scarcity. Since by tradition older fields are usually given priority in water allocation, it was necessary to provide a compensatory source, even though a formal claim had not been made earlier. The Habu reservoir project, initiated soon after World War II, was therefore designed to provide additional water to the downstream rice-growing area.

The Yahagi multipurpose reservoir was designed to serve a number of water uses in the basin. Besides power generation and flood control, the reservoir is a source of urban water. Some water ($1.54 \text{ m}^3/\text{sec} + 1.44 \text{ m}^3/\text{sec}$) released by the Yahagi reservoir is diverted to the agricultural areas both upstream and downstream of the Meiji and Shidare irrigation systems. The major portion of the developed discharge from the Yahagi dam is supplied to the growing industries and cities situated along the midstream of the river ($6.685 \text{ m}^3/\text{sec}$ for industrial use and $4.43 \text{ m}^3/\text{sec}$ for domestic use). Despite the worsening water scarcity in the Yahagi River basin, some of the industrial water was diverted from the Yahagi dam ($2.67 \text{ m}^3/\text{sec}$) across the basin boundary to the coastal industrial zones located south of Nagoya.

Organizational Arrangements for Water Use Management

As in any developed river basin, water facilities such as dams, headworks, diversion facilities, and canals in the Yahagi River basin have to be operated and managed in a closely synchronized and coordinated fashion, particularly during drought. The need for coordination on a basinwide basis among various water suppliers and users also increases as more and more interconnected water facilities are developed and operated jointly by different agencies.

For this purpose, two organizations were set up in the Yahagi River basin. One is the Yahagi River Water Use Coordination Association established in 1971 with all agencies and organizations concerned with the management and use of water regulated by the Yahagi reservoir as members (see Annex). The role of this association is analyzed briefly below in connection with water use management during periods of drought.[3]

Another organization is the Yahagi River System Comprehensive Water Use Management System established in 1973 when the Habu reservoir began to deliver water to the irrigation areas downstream of the Meiji and Shidare irrigation systems. The management system is designed to regulate water supplies from the Habu reservoir to ensure that the irrigation area will not suffer from severe drought in dry years. It is also supposed to coordinate the withdrawal of water for agricultural, industrial, and domestic uses at two intake gates situated along the Yahagi River. The management system is operated by its secretariat, comprised of the Okazaki Agricultural Land Development Office, Department of Agricultural Land and Forestry Affairs, and Aichi Prefectural Government, with the following functions:

- Review daily the water level at the Habu reservoir and the flow conditions at the intake gates along the Yahagi River on the basis of information provided by field offices.
- Adjust water supplies to the different sectors in light of this information and the daily applications made by different water users to the secretariat of the management system from two days before. The Meiji and Shidare Land Improvement Districts are exempt from this application procedure, but must submit records of water withdrawal of the previous year as well as the plans of water withdrawal for the next year.
- Coordinate the day-to-day operation of water intake gates managed by other organizations such as the Yahagi Dam Management Office under the Ministry of Construction and the field offices of the Chubu Electric Power Co. Ltd.
- Establish liaison and coordinate with all agencies and organizations concerned in times of urgency (see Aichi Prefecture Government 1989:4–5).

Water Use Management Practices During Periods of Drought

In recent years, organizational mechanisms to cope with drought have been established in many river basins of Japan. The mechanisms are designed to facilitate joint actions by water users of the same river system for regulating withdrawals to save water stored in reservoirs. Determining the extent to which withdrawals are to be restricted among different water users is a crucial issue.

According to the River Act of 1964, the priorities of water withdrawal are determined not in accordance with the purposes for which water is used, but in the order water rights were granted. Hence, unless a binding agreement on the joint action against drought is made among licensed water users in the same river system, water users with lower priority rights are not allowed to withdraw water in advance of those with senior rights. Recognizing the social disorder that may arise from rigid enforcement of the water use priorities during severe droughts, it has become necessary to establish a mechanism whereby the needs and interests among different water users can be reconciled.

In view of the extremely tight water demand-supply situation in the Yahagi River basin, the Yahagi River Water Use Coordination Association, as an integral part of its activities, coordinates water withdrawals by different sectors during periods of drought. Table 10.5 shows the restricted withdrawal rates implemented during periods of drought from 1973 to 1987. The table also shows differences in the restriction rates among the agricultural, domestic, and industrial sectors. In general, as is now common in much of Japan, domestic water sector withdrawals are cut back less during times of drought than those of the agricultural and industrial sectors. In this respect, the actual practice of water allocation does not conform to the legally established framework, which favors agriculture.

When the restriction rate of water withdrawal for agricultural use reaches as high as 50 percent, elaborate management of water allocation is necessary to avoid crop damages. The Meiji irrigation system is a good example of successful management. Here, on the basis of a century-long water management experience, the farmers have developed a method to ensure impartial allocation of irrigation water during periods of drought. The method is composed of a two-tier water allocation called the "alternation" and the "major alternation." The former is a system in which the water distribution channels along a single trunk canal are used alternately for irrigation, while the latter is where water supply is alternated between trunk canals (see Meiji Irrigation System 1979:80–84). The relatively effective responses made by the agricultural sector to the repeated incidence of drought in recent years have been possible because the organizations, together with the physical layout of the irrigation systems, enabled the smooth implementation of this time-honored practice.

Due to the paucity of relevant information, it is not entirely clear what specific measures are adopted in the industrial sector to cope with temporary water shortages. Given the high rate of in-plant water recycling, 96.2 percent in 1986, many of the larger firms with water-recycling facilities are probably able to tolerate short-term restrictions of water use as high as 50 percent without significant effects on production.

One final remark may be appropriate regarding the current practice of water saving during periods of drought. In general, the restriction rates

TABLE 10.5   Water Withdrawal Restriction Rates for Water Users Relying on the Yahagi Reservoir During Drought

Year	Period	No. of Days	Restriction Rate by Sector (%) Agriculture	Domestic	Industry	Rainfall (mm) May–Sept.
1973	10 June–23 July	14	20	—	—	764
	24 July–2 Aug.	10	30	—	30	
	3–27 Aug.	25	30	10	50	
1978	11–16 June	6	30	10	30	1,232
	17–21 June	5	50	20	50	
	22–23 June	2	30	10	30	
	31 Aug.–1 Sept.	2	Voluntary restriction			
	2–5 Sept.	4	20	10	20	
	6–11 Sept.	6	55	25	50	
	12 Sept.	1	20	10	20	
1979	20–24 June	5	Voluntary restriction			1,294
	25–28 June	4	30	15	30	
	29 June	1	40	20	40	
1981	9–15 June	7	Voluntary restriction			1,119
	16–25 June	10	20	10	20	
	26–30 June	5	Voluntary restriction			
1982	31 May–30 June	31	Voluntary restriction			1,465
	1–9 July	9	40	20	40	
	10–18 July	9	55	28	55	
	19–26 July	8	55	20	55	
1984	1–13 June	13	Voluntary restriction			963
	14–26 June	13	55	25	55	
1986	7–15 Mar.	9	Voluntary restriction			1,069
	14–29 June	16	Voluntary restriction			
	1–17 Sept.	17	Voluntary restriction			
1987	4–17 Sept.	14	30	10	30	1,011

Source: Aichi Prefecture Government 1989.

of water withdrawal for different sectors are determined on the basis of records of water volume actually used. This practice seems logical and fair, but it may discourage water conservation in established sectors, thereby worsening water shortages in others. For example, it has been reported that irrigation organizations have withdrawn their full

TABLE 10.6   Water Quality at the Selected Monitoring Points in the Yahagi River System

Monitoring Points	Year	Total Nitrogen	Dissolved Oxygen	Chemical Oxygen Demand	Suspended Solids
Standards		>1 ppm	>5 ppm	<6 ppm	<100 ppm
Main course	1973–78 av.	1.05	8.2	2.3	n.a.
of Yahagi	1983	0.45	8.2	2.2	12
	1988	0.60	8.1	2.6	12
Tomoe R.	1973–78 av.	0.96	8.3	2.3	n.a.
	1983	0.44	8.2	2.6	9
	1988	0.54	8.9	2.9	10
Oto R.	1973–78 av.	2.87	5.5	8.4	n.a.
	1983	1.54	7.0	4.8	6
	1988	1.90	6.1	6.2	11
Kanori R.	1973–78 av.	2.81	5.2	5.8	n.a.
	1983	1.47	5.0	5.4	15
	1988	1.59	3.8	5.6	18
Kohda R.	1973–78 av.	2.48	4.8	6.7	n.a.
	1983	2.09	3.5	6.0	9
	1988	2.20	3.1	6.7	10

n.a. = Not available.

allotment of water, even when not necessary for crop yields (Okamoto 1981:86).

Water Quality Deterioration and Its Influence on the Choices of Sources for Irrigation Water

Water quality deterioration is one of the crucial issues that complicates the water demand-supply situation. A case in point, mentioned earlier, is the coastal rice production area (about 7,700 ha) of the Yahagi River. In this area, it has become increasingly difficult in recent years to continue the traditional practice of recycling drainage water from upstream fields to irrigate fields downstream, due to the contamination of the Yahagi tributaries, which serve as the source of irrigation water. Consequently, dependence on the Habu reservoir as an alternative source of water has increased considerably (Tables 10.6 and 10.7).

As pointed out earlier, the exclusive right to withdraw the base flow from the main course of the Yahagi River belongs to the two irrigation systems along the midstream. The coastal rice production area, despite its locational advantage, had therefore relied on water from some of the

TABLE 10.7 Amount of Irrigation Water Withdrawn by Source (River) in the Downstream Agricultural Area (million m³)

Year	Habu-Tomoe	Oto	Kanori	Kohda	Others	Total
Base year						
(1947)	55.6	37.0	17.2	27.8	1.6	139.2
1973	82.8	18.3	24.7	45.9	9.0	180.7
1974	86.7	4.6	24.6	19.4	5.0	140.3
1975	92.1	11.0	27.2	12.0	0.0	142.3
1976	105.4	2.1	29.0	15.7	0.0	152.2
1977	118.5	15.8	37.3	30.6	0.0	202.2
1978	101.2	18.2	15.1	n.a.	0.0	n.a.
1979	109.2	14.7	10.5	32.2	0.0	166.6
1980	139.1	16.6	4.5	27.4	0.0	187.6
1981	117.4	19.9	10.5	12.6	0.0	160.4
1982	111.5	11.9	11.0	14.9	0.0	149.3
1983	139.2	9.1	10.5	13.5	0.0	172.3
1984	130.9	17.8	14.5	29.8	0.0	193.0
1985	158.9	15.9	6.3	17.7	0.0	198.8
1986	135.5	16.8	10.5	33.0	0.0	195.8
1987	135.5	22.1	11.9	37.0	0.0	206.5
1988	152.3	15.0	7.5	23.3	0.0	198.1

n.a. = Not available.

Yahagi tributaries. The limited availability of water, relative to the area to be irrigated, had stimulated farmers to develop fairly intricate systems of water recycling.

Accordingly, when the irrigation project plan for this coastal area was formulated, the original water use scheme was designed in accordance with the following guidelines:

- Use waters of the Yahagi tributaries as the priority source of irrigation;
- Mobilize, to the extent necessary to meet the portion that cannot be provided by that source, the base flow of the Tomoe River (the large tributary on which the Habu reservoir was developed) as a secondary irrigation source; and
- Divert water from the Habu reservoir only to the extent necessary to meet the portion that cannot be supplied by those two sources (Tokai Agricultural Bureau 1983:105).

In the original scheme, the share of withdrawal relative to source was 60 percent from the small streams in the area and 40 percent from the

Tomoe River (including supplementary supplies from the Habu reservoir). However, the latter has increased up to nearly 70 percent in recent years (Table 10.7). This shift in the source of irrigation water itself reflects the contamination of water in the Yahagi tributaries to the extent that it is no longer suitable for irrigation. Thus, water quality deterioration has led to further water scarcity in the area.

In addition to the above, the Yahagi River basin is well known throughout the country for its community-based, basinwide water quality protection movement. The movement was initiated in the late 1960s as a response to the repeated incidence of water contamination-related damage both to crops and fishery resources. It was promoted by farmers and fishermen with a view to protecting the basis for their economic survival. They have mobilized their own organizational and institutional resources, such as the network of farmers' and fishermen's cooperatives as the basis for political bargaining, to ensure that effective measures are taken by the appropriate government agencies at the national, prefectural, and local (municipal) levels. At the same time, they have increased the awareness of local community members. The movement, originally started with the aim of protecting water quality, has broadened its scope of activity in recent years to include land use coordination, particularly in the upper catchment area of Yahagi River, and the revitalization of functional linkages between the upstream and downstream communities. These activities have had a significant bearing on the management of the upper watershed of the Yahagi River (Adachi and Oya 1987).

A Note on Watershed Management

In addition to these two problems, issues involved with watershed management in the Yahagi River basin may deserve special attention. In general, in order to ensure that reservoirs effectively serve the purposes for which they are built, watershed management is essential. The rapid industrialization that took place in the plains of the Yahagi River basin since the late 1960s, and the urbanization that accompanied it, have had a variety of impacts on the upstream mountain areas, including rapid depopulation of the mountain villages. The resultant stagnation of these communities has led to increased difficulty in managing the headwaters of the Yahagi.

One particularly serious consequence is that people in the mountain communities are no longer so much able to invest their time and money in the management of forests. This problem has been addressed in part by establishing a "headwaters management fund" with contributions from downstream water users to assist forest management in the upstream communities. The magnitude of assistance currently extended by the fund is rather limited. What is needed is to further promote the philosophy on which the fund was established and create a system whereby the efforts of the mountain community members in managing the headwaters are properly rewarded.

Concluding Remarks

The Yahagi River basin has undergone rapid socioeconomic changes in recent years. The prime factor that has stimulated the changes is the heavy concentration of automobile-related industries centering around Toyota. Yet, in general, industry does not appear to be a major source of water use conflict. Besides raising watershed issues, this paper concentrated on two types of water use conflict that emerged in the course of regional transformation in the Yahagi River basin, namely:

- Conflict over water allocation among different water users during periods of drought; and
- Reduced range of choice of alternative sources for irrigation due to water quality deterioration.

The former problem may have to be further studied in the light of the overall dynamics of water use and management. More specifically, it would be worthwhile to analyze the problem in relation to the current institutional arrangements for water use management in general, and to the system of water allocation in particular.

The latter problem, like the former, is not necessarily unique to the Yahagi River basin; it is common to many metropolitan river basins. In view of the complexity involved in water quality management, the problem may not be readily solved with the policy tools currently available, at least in the foreseeable future.

Annex:
Yahagi River Water Use Coordination Association

Purpose

The Association is established to contribute to the realization of smooth water management in the Yahagi River system by way of (1) ascertaining in a comprehensive manner the state of water utilization, and (2) consulting among parties concerned the ways and means to promote consistent and rational water utilization.

Agenda for consultation include matters relating to:

- Ascertaining the flow conditions of the Yahagi River system and the state of water utilization;
- Measures for rational water utilization;
- Coordination among water users;
- Establishment of a management system;
- Water quality protection; and
- Close liaison and harmony among parties concerned.

Members

Chubu Regional Bureau of the Ministry of Construction
- Director General of Rivers Department
- River Coordinator

Tokai Region Agricultural Bureau of the Ministry of Agriculture
- Director General of Construction Department

Nagoya Regional Bureau of the Ministry of International Trade and Industry
- Director General, General Affairs Department
- Director General, Public Utility Department

Aichi Prefecture Government
- Director General, Planning Department
- Director General, Public Works Department
- Director General, Agricultural Land and Forest Policy Department
- Director General, Prefectural Public Enterprise Agency

Water Users
- Mayor of Okazaki City
- Chairman of the Board of Directors, Yahagi Yosui Land Improvement District
- Chairman of the Board of Directors, Shidare Yosui Land Improvement District
- Chairman of the Yahagi River Riparian Land Improvement District Federation
- Chubu Electric Power Co. Ltd.
  - Director, Hydroelectric Power Department
  - Director, Land Acquisition and Management Department
  - Director, Works and Operations Department

Notes

1. Estimated in Hanayama and Fuse (1977:6).
2. The annual potential water resources in the dry year of Yahagi river basin is estimated in Tokai Agricultural Bureau (1983), as follows:

Annual precipitation	Average year	1,822 mm
	Drought year	1,377 mm
Potential water resources	Evapotranspiration/percolation	540 mm
	Runoff in a drought year	837 mm
Surface area	River basin	2,264 km²
	Mountainous catchment area	1,600 km² (a)
Annual runoff	River basin	1,900 million m³
	Mountainous catchment area	1,300 million m³

(a) The mountain catchment area accounts for about 70 percent of the surface area of Yahagi River basin.

3. Rules of the Yahagi River Water Use Coordination Association (as of 1980).

References

Adachi, S., and K. Oya. 1987. "The Role of Local Communities in Regional Development and Environmental Management: A Case Study of the Yahagi River Basin, Central Japan," in Vol. 1, *The Japanese Experience: Environmental Management for Local and Regional Development.* Nagoya: UNCRD–UNEP.

Aichi Prefecture Government. Various years. *Aichi-ken Tokei Nenkan* (Aichi Prefecture statistics yearbooks). Nagoya: Aichi Prefecture Government.

Aichi Prefecture Government, Okazaki Agricultural Land Development Office. 1989. *Yahagi-gawa Risui Sogo Kanri Nenpo, 1988* (Yahagi River Water Use Comprehensive Management Bulletin, 1988). Okazaki: Okazaki Agricultural Land Development Office, Aichi Prefecture Government.

Aichi Prefecture Regional Planning Committee. 1989. *Dairokuji Aichi-ken Chiho Keikaku: Aichi-ken 21-seiki Keikaku: Sekaini Hirakareta Miryokuaru Aichi o Mezashite* (The 6th Aichi Prefecture regional plan: Aichi Prefecture 21st century plan for creating attractive Aichi open to the world community). Nagoya: Aichi Prefecture Regional Planning Committee.

Hanayama, Y., and T. Fuse. 1977. *Toshi to Mizushigen: Mizu no Seiji-keizaigaku* (Cities and water: Political economics of water). Tokyo: Kashima Shuppankai.

Meiji Irrigation System Editorial Committee. 1979. *Meiji Yosui Hyakunen-shi* (History of 100 years of the Meiji irrigation system). Anjo: Meiji Irrigation and Land Improvement District.

National Land Agency. 1987. *Zenkoku Sogo Mizushigen Keikaku* (National comprehensive water resources plan: Water plan 2000). Tokyo: National Land Agency.

Okamoto, M. 1981. *Suiriken Mondai no Shuhen* (Issues involved with water use rights). *Jurist*, No. 23:86 (summer).

Tokai Agricultural Bureau, Kiso River System Comprehensive Agricultural Water Use Survey Office. 1983. *Showa 57-nendo Yahagi-gawa Ryuiki Chosa ni Kansuru Hokokusho* (1982 report of the study on agricultural development in the Yahagi River basin). Nagoya: Tokai Agricultural Bureau.

11

Alternative Approaches to Urban Water Management[1]

K. William Easter and James E. Nickum

Of the eight metropolitan areas studied, Madras and Beijing are the most "mature" in the sense that they now face serious recurrent water shortages. Bangkok and Manila are rapidly maturing, due to both increased demands and quality degradation. In the Yahagi River basin of Japan and in Seoul, water shortages are growing primarily because of accelerating pollution problems. Honolulu and Osaka face water shortages only during periods of drought. Water shortages could curtail future urban growth in Honolulu if water cannot be transferred out of agriculture. Osaka is an example of a "post-mature" water economy, where industrial demands in particular have actually declined in the city since the early 1970s due to higher rates of recycling and industrial relocation. Nonetheless, although aggregate demand has remained stable in the Osaka region, including the suburban prefecture, deterioration in water quality may lead to a new mature phase.

In the cases presented in this volume, we find provincial/state and local governments heavily involved in managing urban water systems using largely nonprice allocation mechanisms. Yet it is not clear that the current institutional and organizational arrangements, which have been developed to manage and resolve past conflicts in metropolitan water systems, are best suited for the job in a maturing water economy. In particular, demand management has become more imperative, and with it, qualitative changes in institutions as well as marginal adjustments in current arrangements. These changes range from improved interagency coordination to greater use of markets, to working with water-user groups.

Following the Water

Within a metropolitan water system, there are a number of important transformation points where transactions could or do take place between different entities and, sometimes, different institutional regimes. There are different transformation points in the various stages of project development as well. The central government might plan and authorize a project, but a private firm could build the project and an autonomous utility could operate the completed facility.

The first transformation point involves planning for and obtaining water supply at the source. This may involve installing a pump, building a reservoir, or buying water from existing users such as irrigators. Here, both budget concerns and conflicts with other sectors are important, as are externalities. The transaction costs (Box 1) of source acquisition also depend on the number of entities involved. A subsequent transformation point occurs at the treatment facility. Here, ex ante decisions must be made about the technology to use and the level of treatment to provide. When the facility is in place, finding funds to support plant operation and maintenance becomes an overriding concern.

After the water is treated, it must be delivered to the consumer, usually through underground pipes. Leakage in the delivery system wastes much water and contaminates the rest, especially when supply is intermittent. Replacing pipes is expensive and time consuming, which may involve a number of entities and, where they are allowed, sharply higher water charges. Otherwise the provider must compete with other entities for a share of the public budget to upgrade the pipe system and provide continuous water service. Another important problem at this stage involves unauthorized hook-ups and "stealing" of water, which in turn requires policing and penalties to discourage stealing and nonpayment of water charges.

A final transformation point occurs where water first goes through a meter and then into the household or business firm. Meters are occasionally bypassed or disconnected to keep water charges low. In a number of cases, water levies are just ignored. In these cases, the transaction costs include the cost of policing the system to prevent tampering with water meters, collecting fees, and disconnecting (turning off) water to those not paying their fees. The cost of building a system to disconnect nonpaying users is a transaction-related cost.

As one moves from one part of the water system to another, the agency in charge may change. For example, the water department or utility may be in charge of the water source, such as a reservoir, while the revenue department installs the water meters and collects the fees. Such separation of responsibilities within the water system affects incentives

---

Box 1. Institutions and Transaction Costs

By institutions, we refer to the collection of rules and roles, both formal and informal, by which the economic system and its components are operated. Institutions must be specified, negotiated, enforced, and widely accepted by users. Institutions vary in the degree to which they economize on these and other "transaction costs." The transaction costs associated with resolving water-use conflicts can also be very high. If the costs are higher than the expected benefits of establishing an appropriate institutional structure to resolve them, conflicts may not be resolved until they become serious.

High transaction costs that perpetuate externalities are found in several different situations. For instance, many countries have weak judicial and enforcement systems. Furthermore, establishing property rights is itself expensive. As water becomes more scarce and therefore more valuable, it may become more worthwhile to specify and enforce property rights. On the other hand, the complexities of incorporating the interests of a growing number of stakeholders may make it more difficult to establish such rights.

A transaction-cost-based analysis of most of the urban areas in this volume may be found in Nickum (1991).

---

and raises transaction costs. Entities with responsibility for water delivery may lack incentives to provide good service unless they also collect and depend on water fees to support their activities. Autonomous entities that must cover costs from collecting water charges have a strong incentive to provide good service because satisfied customers are much easier to collect fees from than dissatisfied ones.

Although transaction costs are critical in determining the water system that best suits a given metropolitan area, other factors are also important, such as production costs and economies of scale. The question is, What management strategies should be adopted that best takes into account the above factors at various points in the water system?

Management of Maturing Systems

Since many of the water supply systems in Asia, as elsewhere, have been in the expansionary phase of water development until recently, the principal response to water shortages and water-use conflicts has been to build another project designed to deliver sufficient amounts of water to meet the "need" of all competing uses. The planning mechanism by which this is done is to estimate demands for a future year (e.g., 2020), based largely on past usage coefficients, and then select supply options that can meet the estimated gap. This is termed "supply management," in contrast to "demand management," which are activities targeted at reducing the use rates.

Beijing provides an example of supply expansion to resolve or avoid a conflict over water use. When Beijing Municipality claimed the exclusive right to use water from the Miyun Reservoir, which had been shared with Tianjin Municipality downstream, the Chinese central government built Tianjin a new reservoir outside Beijing's watershed (see Chapter 3 by Nickum, this volume). In Japan, agriculture claims the water rights (*suiri ken*) to almost all the base flow of rivers that may be withdrawn. Hence, increments to industrial and urban water supply, especially during the high growth of the 1950s and 1960s, were provided through the construction of reservoirs that tap the incremental flow above base levels (see Chapter 6 by Akiyama and Nakamura, this volume).

Besides easing conflicts among users, supply-oriented management can take advantage of economies of scale, can hold down transaction costs, and is relatively easy for a government agency to implement, particularly if it is construction-oriented. But supply management is becoming more expensive (Box 2), and the costs are often borne, in large part, by non-beneficiaries (i.e., the general taxpayer). As the water economy matures, new water sources become more expensive to develop, and conflicts increase. Hence, the focus of water policy must shift from opening up new supplies to demand-oriented management strategies, including the reallocation of existing supplies.

As new sources of water become increasingly expensive and difficult to finance, the transaction costs involved in demand management strategies will not appear to be so high. Demand management tools, such as higher water charges, water conservation technologies, rationing, and water quotas, are already being more widely used, in part because of improvements in monitoring and billing capabilities. Of these tools, a pricing approach is especially attractive in the long run. Besides reducing use rates, higher water charges produce at least two benefits. First, they usually generate more revenue, providing water agencies with additional funds to develop new water supplies or to reduce water losses in the existing system. Second, they show more clearly how much more valuable an additional (marginal) unit of water is when it is used for domestic or industrial uses rather than to grow wheat or rice. In contrast, rationing or quotas are generally used during drought periods and do not generate any added revenue or information on the value of water. Once a metropolitan area reaches a stage where water use can be analyzed in a supply-demand framework, an optimal strategy can be selected based on the costs (including transaction costs) of various supply and demand management alternatives.

Box 2. Increasing Costs of Water Supply

Many cities convey water over long distances and make extensive use of high-cost pumping. Furthermore, additional water treatment has been increasingly necessary.

Shenyang (China): The cost of new water supplies would rise between 1988 and 2000 from $0.04 to $0.11/m³, almost a 200 percent increase. The main reason is that groundwater from the Hun Valley Alluvium, the current water source, had to be discontinued as a source of potable water because of its poor quality. As a result, water will have to be conveyed to Shenyang by gravity from a surface source 51 km from the city. In Yingkou, another Chinese city, the average incremental cost (AIC) of water diverted from the nearby Daliao River is about $0.16/m³. However, because of heavy pollution, this source cannot be used for domestic purposes. As a result, water is currently being transported into the city from the more distant Bi Liu River at $0.30/m³.

Amman (Jordan): When the water supply system was based on groundwater, the AIC was estimated at $0.41/cm³, but chronic shortages of groundwater led to the use of surface water sources. This raised the AIC to $1.33/m³. The most recent works involve pumping water 1,200 m from a site about 40 km from the city. The next scheme contemplates the construction of a dam and a conveyor, at an estimated cost of $1.50/m3, a cost comparable to that of desalinating seawater of $1 to $2/cm³.

Lima (Peru): During 1981, the AIC of a project to meet short- to medium-term needs, based in part both on a surface source from the Rimac River and on groundwater supplies, was $0.25/cm³. Since the aquifer has been severely depleted, groundwater sources cannot be used any longer to satisfy needs beyond the early 1990s. To meet long-term urban needs, a transfer of water from the Atlantic watershed is being planned, the AIC of which has been estimated at $0.53/cm³.

Mexico City: Water is currently being pumped over an elevation of 1,000 m into the valley of Mexico from Cutzamala River through a pipeline of about 180 km. The AIC of water from this source is $0.82/m³, almost 55 percent more expensive than the previous source, the Mexico Valley aquifer. The former source has been restricted due to problems of land subsidence, lowering of the water table, and deterioration in water quality. The newly designed water supply project for the city is expected to be even more costly, since it will have a longer transmission line, and water will be pumped over an elevation of 2,000 m to the city.

Note: Costs exclude treatment and distribution.
Source: World Bank 1993.

Management alternatives depend on a country's hydrologic stage of development and the age of its metropolitan water systems as well as the stage of its economic development. Old systems in Third World countries pose special problems of the high cost involved in maintaining and upgrading their water systems when faced with chronic capital shortages. An equally difficult set of problems arises when metropolitan areas find that some of their fully exploited low-cost water sources may not be available in the future (e.g., due to groundwater pollution or because others have senior water rights that are not transferable).

Institutional Alternatives for Managing Water Supplies

An increased emphasis on demand management immediately raises the issue of how water is to be managed, and in particular, how conflicts over use are to be resolved. There are three general institutional forms: the market, administrative (command-and-control), and collective. Market-based approaches to allocate water are increasingly popular in countries such as the United States, the United Kingdom (except Scotland), and Australia. Direct water management by government agencies or public utilities is most widely used in developing as well as in many developed countries. Conflicts over water use are then either resolved by these agencies or their higher authorities, or find expression in interagency conflict or supply management solutions. A third approach is to rely on nongovernmental, often community-based, water-user groups.

These three approaches for resolving water-use conflicts are not mutually exclusive; in fact, they may reinforce one another. For example, a strong state adjudication and enforcement system may lower the transaction costs of establishing, defending, and transferring water rights sufficiently to allow private, market-based allocations of water. Otherwise, an alternative system of water rights may prove to be more efficient in allocating water. Ironically, this is unlikely to be a system of state ownership and/or state use rights, since a government that cannot adjudicate effectively is unlikely to allocate effectively. It is more likely to be one where water management rights are handled by water users or their community (e.g., village or ward).

Market-Based Approaches

The role of the market as a decentralized dispute resolution device has long been recognized by economists (e.g., Grossman 1974:64). Although markets are associated with private ownership, they are not necessarily so. The key ingredients are not ownership but competition, or "contestability," and freedom of choice, especially by the buyer. Hence the use by state agencies of contestable service contracts or construction contracts with private firms can improve the quality of service and reduce costs by introducing competition.

Markets for irrigation water have been discussed intermittently for years. They have worked efficiently to transfer water from agricultural to municipal uses in some parts of the American southwest (Saliba 1987). Nonetheless, even where markets exist, it has been difficult to develop institutional arrangements for exchanging water or water rights. The characteristics of water that make this so include the pervasive interdependency among users, the "public good" nature of many water sources, economies of scale in large water projects, high variability in supply and demand, and conflicting social values concerning water. In addition, the cost of making transactions is high when a large number of water users are involved. For large systems, communications and information problems make it difficult to allocate water efficiently and equitably to all through markets or government-administered systems.

Another problem is the availability of economic rents (above-normal returns) to water utilities and alternative suppliers, restricting the area served by subsidized supply. For example, a water vendor in Jakarta charges substantially higher prices for water than would be necessary to earn the average market wage. It is not clear, however, how much of this is absorbed in payoffs to neighborhood officials and water utility staff (Lovei and Whittington 1991). Capturing these rents depends partly on economies of scale and the level of subsidy, and on the artificial imposition of transaction costs on alternative sources of supply. Lovei and Whittington recommend the deregulation of private water sales, including sales to connected households, to lower artificially created transaction costs.

In the past, many water development activities for both the urban and rural sectors have involved large government subsidies. In some cases water charges were low because the primary purpose of the development is to produce a public good or service, such as hygienically clean water. In others, the difficulty lay in measuring water delivered and in collecting water fees. However, these problems are much less severe in a metropolitan setting than in irrigation.

Finally, government subsidies, both to justify their existence and for them to function, must be accompanied by restrictions on their use (e.g., eligibility requirements). One of these restrictions is usually a prohibition on resale, which clearly inhibits the market mechanism.

Tradable Permits

When water rights, represented by water permits, can be bought and sold, water will naturally be reallocated to uses with higher economic value if transaction costs are low. The prerequisites for keeping transaction costs low are (1) an adequate information system, (2) a delivery system that allows for transport of water sold to the buyer, and (3) an organizational arrangement for implementing the reallocation required by the trade. In addition, the rules for water delivery must be sensitive to changes in conditions so that trades or sales do not require extensive bargaining, which raises transaction costs (Easter 1986). Conveyance structures and organizational arrangements must be able to accommodate changes in location and type of use. A major problem is that permits may be unable to account for interdependencies among water users (return flows).

Contracting for Service

Two areas where market-like forces could be used are the provision of services for public utilities or government agencies and the actual operation of water companies. For example, the urban water sector in Côte d'Ivoire (Ivory Coast) has been operated for the past 25 years by a private company under a mixture of concessions and lease contracts. This operation is considered one of the best-run utilities in Africa. In Macao, the percentage of unaccounted-for water fell by 50 percent over 6 years following privatization of water supply services in 1985. Similarly Guinea, which recently promulgated a lease contract for supplying water to its principal cities, experienced dramatic improvements in its earnings. Within the first 18 months, bill collections increased from 15 to 70 percent (*World Development Report* 1992:111).

Santiago's water utility successfully contracted out to private entities its services in meter reading, pipe maintenance, and billing. Productivity of the utility's staff rose three to six times higher than their counterparts elsewhere in the region. Other Latin American cities such as Buenos Aires and Caracas have begun to use concessional service contracts (*World Development Report* 1992:111).

Indonesia recently tried a concessionaire contract for water resource development in East Java. The main component of the contract was to build a 65-km pipeline. Several groups expressed interest; however, the Bromo consortium, backed by local and international financing, won the contract. The agreement includes building the pipeline and operating it for a minimum of 15 years.

Greater use of market-like approaches is not without problems. Economies of scale make it difficult to have direct competition among suppliers in a specific area. Furthermore, finding private sector contractors is often difficult, particularly in developing countries. Even internationally, often only a few firms compete for such contracts (*World Development Report* 1992:111).

Government-Based Nonmarket Approaches

As noted earlier, nonmarket administrative (hierarchical) approaches such as "command-and-control" are not synonymous with government. Public agencies can use market methods, while private parties can run hierarchies, especially in governing intrafirm transactions. Here we focus on government nonmarket approaches, however.

Government-based administrative approaches correct some market failures, but they are prone to "nonmarket failures" that can significantly raise transaction costs, particularly in rigidifying existing institutional or organizational arrangements. These failures include conflicts between and among administrative units, rent-seeking behavior, redundant and rising costs (without competition there may be little incentive to improve efficiency), unanticipated side effects (e.g., pollution), and distributional inequity based on political power (Wolf 1979).

One of the most important reasons for government failure has been the lack of accountability of government agency personnel. In the Philippines and elsewhere, some accountability has been achieved by requiring irrigation officials to recover costs directly by collecting fees from farmers. This requirement gives farmers a means to show their displeasure at poor service by withholding payment. In turn, this action motivates irrigation personnel to see that the system is efficiently operated and maintained because their job security and salaries are at stake.

Governments and Markets

Although a government may install a metropolitan water system and set water quality standards, it does not have to operate the system. As pointed out earlier, governments can contract with a private firm to manage the water system or even sell the system, as the British have done. In the Asia-Pacific region, however, the trend has been the opposite: private supplies are supplanted by the public, due to the economies of scale of the latter (e.g., the areawide systems in Osaka and Korea) and to the open access problems of the former, especially in the exploitation of groundwater (e.g., in Osaka and Bangkok). In the region, where private suppliers gain in importance, it is usually due to the opening up of new supplies (groundwater or rivers outside the domain of the state agency) rather than a transfer of authority from the public.

Legal Structure

Even if markets were to be legitimized, government involvement would be necessary to make an initial assignment of water rights (and the duties such as payment of taxes and fees attached to them). Furthermore the way these rights are designed can facilitate or hinder market transactions. Historically, groundwater rights have been tied to land, and transfers of rights have required the transfer of land. Surface water rights are easier to separate from the land, but in many cases, such as under the "riparian rights" doctrine, they are also tied to land. This acts as an unnecessary constraint to water transfers.

The government can attach a number of conditions to water rights (e.g., that one must put the water to beneficial use or that streamflows must be maintained at a certain level). Governments also specify priorities in case of scarcity—either in terms of type of use (agricultural, industrial, or municipal) or seniority of rights (first in time, first in right).

Randall (1981:202) states:

> The rights transmitted by negotiable entitlements must be specified in terms of the secure expectations of the rights holder and the duties and obligations of the water authorities. Particular issues requiring resolution include:
>
> - the time-span of the entitlement and provisions for rental rights to deliveries in the event that long-term entitlements are specified;
> - the method of accommodating the stochastic nature of water;
> - availability (possibilities include individual rights to some specified fraction of water available for delivery, and the specification of different classes of entitlements in terms of reliability, i.e., the probability of water delivery);
> - the time and place of delivery;
> - the ownership of tailwaters and return flows and the attendant obligations upon the owner; and
> - the conditions under which entitlements could be transferred, with special reference to transfers which would change the time and/or location of water demand.

Pollution Control and Water Supply

A wide range of policy tools is available for addressing water quality problems, including fiscal incentives, land use management (retirements, easements, and zoning), outright bans, standards, tradable permits, and strict liability for polluters. Government also makes direct financial commitments such as loans for building water treatment plants, sewage facilities, and major drains. The degree to which governments actually control water pollution varies considerably, depending on factors such as hydrogeological conditions, the level of the economy and fiscal conditions, and the desire for clean water by decision-makers.

Although tradable permits are increasingly common for the right to pollute the air, especially in the United States, attempts to use them to control water quality have not been successful to date (Letson 1992). It has apparently been difficult to draft a set of regulations that provide adequate safeguards for the public while providing polluters an incentive to make a trade.

Complicating the problem of control is our inadequate understanding of risks to human health posed by various pollutants, such as agricultural chemicals in drinking water. Consequently, one of the key problems in implementing a pollution control program is inadequate information (Nickum 1993). Water pollutants are of many types and have many sources. We generally do not know what the effects (the "dose-response relationship") and the total exposure of potential victims are. If an individual consumes small amounts of atrazine, how much does his or her risk of cancer increase? What difference does it make whether the person consumes the same amount of atrazine in a few large doses or in smaller quantities over an extended period? Does the age or health of the consumer affect susceptibility?

An additional problem is the ability to detect and measure pollutants. The perceived quality of Honolulu's water has declined as the ability to measure ever smaller concentrations of chemicals has increased. Changes in measurement capabilities, perception of hazard, and budgets of responsible organizations also yield spotty records of water quality monitoring.

Industrial pollution control is complicated, not only because of uncertainties in the dose-response relationships but also due to the rapid proliferation of new chemical compounds, many of them virtually indestructible. Agricultural pollution shares many of these complexities and is geographically more widespread, with more numerous sources. When the number of sources becomes too large to measure (or control) directly, pollutants are said to have a "non-point source." Agricultural pollution is usually of the non-point source type, as is runoff of automotive wastes from streets and highways and, frequently, urban domestic waste such as discarded motor oil. By definition, non-point sources of pollution are difficult to identify or control.

Although point sources are much easier to monitor, it is sometimes difficult to control them, largely for institutional reasons. The major point sources tend to be large industrial facilities and municipal sewage plants. In the former case, the institutional problem is related to goal conflict: economic growth (narrowly defined) versus the environment. Powerful private interests may exert their influence on government officials to protect the rents (above-normal profits) they are extracting from the environment and society at large. This problem is not solved through public ownership, as then the rule-maker, the state, is also the principal rent-taker.

Moreover, a government agency may find it more expeditious not to enforce environmental regulations. For example, in the United States, the National Pretreatment Program of the Clean Water Act requires industries discharging large amounts of toxic materials and other wastes to treat them before they are discharged. Yet municipal treatment plant managers are reluctant to act against violators. They choose, in effect, to deal with the conflict through avoidance, although it may mean damage to plant facilities and be detrimental to workers' health (United States General Accounting Office 1989:19–20).

We see this problem from a slightly different angle when we look at municipal sewage treatment. If a city is required to construct a new sewage treatment plant to meet water pollution standards, it will usually face a budgetary problem, even in higher income countries. It is usually difficult to increase revenues for this purpose through taxes, and borrowing such a large sum may obligate the city without directly generating new revenue. Thus, without considerable external subsidies, new sewage plants are not built.

The cost of capital is not the only or even the primary inhibitor to controlling water pollution. In fact, the level and nature of the substantial costs involved in obtaining information, monitoring, enforcing, and administering different control strategies can dictate which policy tool the government selects. These costs are related closely to opportunity costs—the availability of alternative technologies and resources. For example, if good, relatively safe substitutes are available for a highly toxic pesticide, the enforcement costs of an outright ban may be quite low. On the other hand, it will be difficult and even impossible to enforce a ban without substitutes.

Different pollution control tools provide different mixes of deterrence, compensation, and information. An outright ban with severe penalties acts as a strong deterrent, both to deviant behavior and to the revealing of information by those who violate the ban. Conversely, standards without significant penalties are not as effective in ensuring compliance, but may encourage the sharing of information.

In many cases, there are substantial economies of scale in constructing, operating, and maintaining waste disposal, water treatment, and water supply facilities. These economies may call for joint ownership of water facilities and building interconnections between major sources of water that serve a number of communities. Joint ownership, along with system interconnections, makes it much easier to take advantage of economies of scale and to move water to the areas of highest demand. Both are probably required, since without joint ownership, interconnections may soon be broken, such as the case in New Jersey.

During World War II, the U.S. War Production Board issued a directive which required the prompt interconnection of all water supplies in New Jersey, a heavy war-industry state. After the war, most of those interconnections were just as promptly severed. Thus, when a deep drought gripped the Northeast in the early 1980s, Abel Wolman, then a consultant to the State of New Jersey, reminded the autonomous, independent water-short communities of those interconnections that could have been used to transfer water to drought-stricken jurisdictions (Revelle 1988:2).

The New Jersey case also shows that it may take some outside government agency to implement an efficient solution to the water problems of local communities. Many times, independent communities reject water supply or treatment solutions that would benefit everyone because they are afraid of losing some independence or they feel that their cost-share is too high. Under such conditions, national or state governments need to use a combination of incentives and restrictions to encourage cooperation among communities.

In the regions discussed (this volume), Osaka and Korea have successfully developed areawide networks serving a large number of municipalities. So has southern California.

The purity of publicly supplied water is excessively high for many of its actual uses. One reason for this is the cost of creating multiple delivery and disposal systems for water of different qualities. Where separate delivery systems are installed for less demanding uses (e.g., for irrigation water or industrial water), purification costs can be lowered.

Administrative Pricing

Particularly in the provision of domestic water supplies or waste disposal services, there is a good opportunity for pricing by volume used. In many cities, such as Beijing, each house or apartment block has its own water meter. There the problem is not lack of meters, but the pricing structure, failure to collect, and in some cases, damaging or circumventing of meters with impunity. Mostly the water charge is set to cover some of the recurrent costs of operation and maintenance. In all too many cases, a replacement charge for facilities is not included. At best, the charge is volumetric. At worst, there is a fixed charge for water, no matter how much is consumed, or a declining rate where a lower price is assessed for each additional unit of water used. With a few exceptions, such as Beijing and Istanbul (Box 3), municipalities have not used an increasing block rate, where the unit price goes up with increases in water use. Usually, higher prices are not even charged during the summer or other peak periods to discourage water use.

Box 3. Increasing Block Volumetric Water Charges in Istanbul

Water and sewerage in Istanbul are managed by an autonomous agency, the Istanbul Water and Sewerage Authority (ISKI). Only limited subsidized financing is available for water and sewerage investments from central government agencies and the municipality. The ISKI has been obliged, therefore, to finance most of the investment program, as well as the operation of the system, from charges to customers. The issue has been to raise the necessary funding, while assuring that the basic water needs of the population are met, including, in particular, those of lower income households (about 18 percent of the population is estimated to have incomes below the poverty threshold). ISKI has used the practice of increasing block volumetric water charges for households to meet these dual objectives. Charges were set in 1987 as follows:

Customer category	Basic rate (US$/$m^3$)	Consumption (%)
Households		
0–7.5 m^3/month	0.26	4
Above 7.5 m^3/month	0.53	45
Office	0.88	12
Industry	1.24	38
Average	0.68	

Households are thus cross-subsidized by industry: water charges up to a minimum to meet basic health requirements (7.5 m^3/month), which are kept particularly modest. Affordability calculations indicate that low-income households would spend about 2.7 percent of average monthly income on water and sewerage charges, compared with 3.4 percent of median-income households. Water and sewerage charges have been raised several times since 1987, and the principle of full payment by consumers for investment and operating costs has been accepted. This system has worked because consumers are metered, and vandalism of meters has not been a serious problem. Still ISKI faces major problems in billing and collection, with billed water amounting to about 65 percent of water consumed. ISKI has also faced rising investment costs, and sites originally planned for sewage treatment have been found unacceptable by local communities. Despite these difficulties, the ISKI model has been adopted by four other large municipalities in Turkey, all of which have introduced the increasing block volumetric water charges.

Part of the problem is the lack of information, this time of an economic nature. We do not know how people will react to higher water prices. For some uses, we would expect that the price elasticity would be quite low, while for other users it may be fairly high. If we are dealing with domestic uses that are price inelastic, then water prices will have limited impact on use. In such cases, water prices or charges will be mostly a means to generate revenue. However, during the summer, consumers appear to be fairly responsive to calls to conserve water, at least in developed countries. Thus, high prices during such peak periods will encourage reductions in water use.

Another reason water or sewer charges are not used to encourage conservation is political. Many local users consider water and sewer charges as just another tax instead of a price for a good or service. They complain about increasing water charges even when they are provided better service, particularly if they are large consumers. Furthermore, many city administrators are often afraid that high water and sewer rates may discourage industrial development. Yet it is not at all clear that the level of water and sewer rates has much effect on the location or expansion of most industries.

Inaction

The Chinese Taoist (Daoist) philosopher Laozi anticipated Adam Smith by over two thousand years in counseling kings that the best way to rule was often not to intervene in the natural course of events (*wuwei*). Where conflict resolution mechanisms exist in society, government involvement may be redundant or even an impediment. For example, if there is a possibility of a subsidized, supply-oriented government solution to a local conflict over water use, water users are less likely to work out an accommodation among themselves that would require them to reduce their water use. If government resources for support are limited, as they usually are, water users (including local levels of government) will often prefer to wait their turn or devote their resources to lobbying efforts. Thus government involvement can actually discourage private sector investment and collective development by users. In addition, non-involvement that stems from political paralysis or instability affecting regulatory regimes can also discourage investment.

Collective Action by Users: Important to All Solutions

Collective action can be useful in resolving externalities and other problems previously mentioned. The idea is not new. As Swaminathan (1986:v) aptly states:

People dependent upon renewable natural resources have evolved ways of managing them properly. When they have failed to do so, the people, the resources, or both have disappeared. Communities have developed such institutionalized forms of control as irrigation councils in southern Asia, forest-cutting controls in Nepal, wildlife utilization taboos and regulations in the Congo Basin, the hema system of pasture protection in Arabia, fishermen's indigenous associations in western and southern Asia, and land use management for conservation in Zimbabwe.

Although such excellent examples of collective action exist, it has not been easy for governments to initiate sustainable collective or community activity, except under conditions where (1) communities were relatively small, stable, and homogeneous, (2) community leadership was strong and representative, (3) benefits from cooperation were high and relatively evenly distributed, and (4) the communities had experience in providing collective goods (Easter et al. 1986; Easter and Palanisami 1976). Examples abound of inactive water user organizations that governments tried to establish. However, progress is being made in some countries in establishing community-operated water systems.

Basic Principles: Assurance, Reciprocity, and Fairness

Individuals will generally not contribute to a community project or abide by resource conservation rules, unless they are reasonably sure that others will do likewise. This is known as the expectation of "positive reciprocity." Runge (1984) found the key to cooperative behavior in the development of institutions that provide such assurance. Ostrom (1990) has derived similar conclusions from repeated game simulations and a range of empirical examples.

Urban-rural water conflicts are analogous to watershed protection situations in that the problem of reciprocity is complicated by the separation of costs and benefits. Often these conflicts show up as between upstream (rural) and downstream (urban) users. People in the upper watershed are tempted to neglect that part of their soil conservation efforts that would only produce benefits for downstream water users (Dani 1986). A number of alternatives have been tried to overcome this problem. Before 1920, in Japan "irrigation associations and municipalities downstream were very active in improving the deteriorated watersheds at their own expense" (Kumazaki 1982:113). More recently, the Yahagi River Riparian Water Quality Preservation Policy Association, centered in the Meiji Irrigation Association, has forged reciprocal links of understanding and economic benefit between upstream and downstream users in the Yahagi River basin (see Oya and Aoyama, Chapter 10, this volume).

Thus, collective action and cost-sharing by downstream water users can be an important way of improving water quality and quantity. One of the key components of such collective action is a good understanding by downstream water users of the benefits they receive from conservation activities upstream. Also, institutional arrangements that allow them to assist in conservation activities need to be in place. If they are cost-sharing, then they need to know that the funds will be efficiently used for desired purposes. Cost-sharing should be considered fair by upstream land owners, which would encourage them to engage in more conservation practices.

As indicated in the Japanese example, such cost-sharing may be arranged directly by the parties concerned, especially when the number involved is relatively small. As the economy develops, however, the demands on water increase and the number of interested parties involved may grow as well. In that case, cost-sharing may be mediated by a higher authority, usually the government. As the water economy continues to mature, the problems of direct coordination are likely to exceed the capabilities of individual government agencies as well and will require coordinating mechanisms such as a river basin commission or a national water council.

Establishing Water User Organizations

Establishing water user organizations (WUOs) requires government commitment and facilitating institutions to implement decentralized water management, plus the proper conditions at the local level. Most of the literature on WUOs has focused on irrigation and rural water supply. We will review the lessons from that experience before assessing its applicability to urban areas.

Some Lessons from Irrigation WUOs. Two factors stand out as critical to the. success or failure of an irrigation WUO: proper incentives and solid, accountable leadership.

- *Incentives.* Hunt (1985) argues that the lack of economic incentives has caused many WUOs established by outside initiative to be ineffective.

  > It seems obvious that WUOs will not produce bottom-up leaders and the farmers will not do the dirty work without access to sufficient rewards. The principal reward to be gathered from the whole process is real control over a predictable delivery of water. And this incentive, for so I believe it to be, will be difficult to achieve without the farmers themselves having strong control over the distribution of water (Hunt 1985:30–31).

Coward (1986) goes even further to argue that local ownership (of irrigation facilities in this case) is necessary to provide the incentives for local participation in project operation and maintenance.

> If the state wishes to have cultivators act jointly to operate and maintain some or all of the hydraulic facilities which they use, it appears that group action based on property relationships will be required. The state can invest in a manner compatible with those principles by investing indirectly. That is, it should provide subsidies and other inputs which complement local investments but also ensure that the hydraulic property which is created is owned by those responsible for its continuance (Coward 1986:242).

The Philippine government has made a serious attempt to develop such incentives in its small irrigation projects. By the end of 1983, the National Irrigation Administration (NIA) in the Philippines had organized over one thousand WUOs. NIA gave many of them responsibility for operation and maintenance. (For a useful description of NIA's "learning process," see Korten and Siy, 1988.) The participatory approach has also involved water users at the construction or rehabilitation stage. This, combined with WUO responsibility for repaying a portion of project costs, has had a positive impact, as farmers then showed a keen interest in putting the water to its best use.

- *Leadership.* In a study of small irrigation systems in northeastern Thailand, leadership was identified as a key variable in explaining differences in the success of irrigation projects (Easter 1986). The Yahagi experience (Chapter 10, this volume), which has depended critically on the efforts of one key person, Mr. Renzo Naito, confirms this. Both the Philippines and Sri Lanka experimented with institutionalizing leadership in the form of government-sponsored community organizers. These organizers, in combination with a program of leadership training, agency reorienting, technical assistance, and the shifting of responsibility for water allocation to WUOs, appear to offer one of the best approaches to improving local participation.

  Accountability of WUO leadership, particularly financial accountability, is also important for the continued success of local water management. In the Philippines and Thailand, one of the most frequently cited reasons for failure of community or communal associations is financial mismanagement.

- *Other Factors.* Vermillion (1991) presents a number of preconditions for an effective self-managed irrigation institution. These include a clear definition of system boundaries and service access rights; benefits of investing in irrigation institutions in excess of opportunity costs; a practical and accountable system of monitoring and regulating behavior; and accessible and low-cost conflict resolution arrangements. To these we would add (1) a relatively homogeneous membership (to facilitate assurance, reciprocity, and fairness); (2) transparent cost-sharing arrangements, which make the most economical alternative attractive to participants; and (3) cultural values, which stress cooperation over conflict.

Rural Water Supply. Programs to develop rural water supplies with community participation are widespread in many developing countries, including Colombia, Malawi, Paraguay, Bolivia, Kenya, and Bangladesh. Colombia has the reputation of having the best rural water supply program in Latin America. The National Institute of Public Health encourages community-managed systems by providing design standards, instruction materials, technical assistance for maintenance, and organizational assistance. The community participates in project designs, elects the administrative committee, raises funds through social activities, and provides material, labor, transport, and cash for construction. An administrative committee operates, maintains, and regulates each system.

Similarly, the program to develop water supplies in rural Malawi promotes community organization and provides design standards and technical assistance. The program has been quite successful, starting with a community of 2,000 and expanding to communities throughout the nation. Community owned, maintained, and operated rural water systems provide over a million people with safe, reliable, and convenient water supplies.

The programs in Bolivia, Kenya, and Bangladesh all recognize the essential role of women in operating community water and sanitation facilities. Because there is a tendency to ignore women when those facilities are planned and installed, a determined effort was needed to ensure their involvement at all stages. The result was successful projects that are effectively managed by local communities with active involvement of women. For example, by 1988, 135 village committees existed in Kenya, all of which had women as treasurers and all had functioning pumps. In the project area between 1985 and 1987, there was a 50 percent decline in diarrhea and a 70 percent decline in skin diseases.

WUOs for Urban Areas? It is increasingly appreciated from experience that formal or informal community-based water user or sewerage organizations can accomplish a wide range of management tasks. Furthermore, they can provide a key ingredient to conflict identification and resolution, even where market or agency-based approaches prevail.

Many of the water user organizations listed in the literature deal with irrigation, village water supply, or urban sewage systems and have a relatively small membership. An open question is, How do we apply the lessons learned from such groups to the wider issues of intersectoral water conflicts in large Asia-Pacific metropolises, sometimes involving millions of people. Can (or should) urban WUOs be set up to supply urban household uses? Is a metropolitan water district the appropriate urban counterpart to an irrigation WUO? How do urban WUOs interact with rural ones, if at all?

Similar questions may be posed by government agencies. Can the experience of irrigation agencies such as in the Philippines be usefully applied to other government organizations involved in water use? What is the mode of interagency interaction? What is the appropriate scale of government organization to properly regulate the water supply and demand? Is a multipurpose authority called for at the national or river basin level?

Some interesting developments may point to an answer. One is the rise of basin-wide, user-based management bodies such as the Yahagi River Riparian Water Quality Preservation Policy Association, established in 1969. Also, despite the tendency of some outside analysts and donors who regard irrigation users' organizations as being irrigation-specific, irrigation associations commonly engage in other activities. Finally, there has recently been a dramatic increase in the number and effectiveness of grassroots environmentalist groups in many developed and developing countries.

Urban Community Sewerage Systems. Innovative technology in Karachi squatter settlements and in northeast Brazil, combined with community organizations, has resulted in the successful provision of community sewerage systems. In northeastern Brazil, condominial systems provide services for thousands of urban people. The individual households are responsible for maintaining the feeder sewers while the utility is responsible for the trunk mains. This division of responsibilities has created several important incentives.

> First it increases the communities' sense of responsibility for the systems. Second, the misuse of any portion of the feeder system . . . soon shows up as a blockage in the neighbor's portions of the sewer. The consequence is rapid, direct, and informal feedback to the misuser. . . . And third, because of the greatly reduced responsibility of the utility, operating costs are much lower (*World Development Report* 1992:07).

In Karachi, with a small amount of external funding, the Orangi Pilot Project (OPP) was started to provide sewage to squatter settlements. The project (1) reduced the cost to affordable levels and (2) developed organizations that could provide and operate the systems. The OPP staff played a catalytic role in establishing the organizations.

> The households' responsibilities include financing their share of the cost, participating in construction, and electing "lane managers" who typically represent about fifteen households. Lane committees, in turn, elect members of neighborhood committees (typically representing about 600 houses) which manage the secondary sewers (*World Development Report* 1992:109).

The OPP has provided sewage services to more than 600,000 poor people in Karachi. Recent initiatives by several other municipalities in Pakistan have followed the OPP method: "Even in Karachi the Mayor now formally accepts the principle of 'internal' development by the residents and 'external' development (including trunk sewers and treatment) by the municipality" (*World Development Report* 1992:109).

Conclusions and Lessons Learned

Demand management strategies have been used only to a limited extent in the eight study areas. Water charges are primarily a means to collect revenue and pay operating costs. The future of supply management approaches does not look promising for many of the areas since alternative sources of supply are becoming prohibitive in both production and transaction costs. Thus, these metropolitan areas must do a much better job in the future of allocating their existing water supplies. Water pricing, conservation, and reallocation promise to be at the forefront of future policy options.

The most effective approach to resolving conflicts and providing water to consumers is likely to vary across countries and by type of service. In this context, separating management of water resources from delivery service is useful. These two broad functions present quite different management challenges and potentials for market failure. At the water *resource* management level, interdependencies and externalities are critical, as is the long-term lumpy nature of many water investments. For delivery of *services*, the problems of monopoly pricing and the highly visible nature of water supply and demand are more important. These characteristics suggest that, in general, governments need to manage water resource development while water user groups and the private sector can play a more important role in the delivery of services.

The management of water resources is defined to include overall resource planning, securing water supplies (new reservoirs, water diversions, establishing water rights), and investing in major water facilities

Box 4. High Cost of Water for Urban Poor

In 1988, over 130 million of the developing world's poorest people lived in urban areas. About two-thirds of these people live in squatter settlements. Most of them depend on traditional sources of water supplies that are becoming increasingly contaminated by human waste, industrial effluent, and agricultural pollutants. As a consequence, the poorest have to purchase safe household water at high prices for meeting their basic needs.

In a number of studies, the urban poor have been shown to pay high prices for water supplies and consequently to spend a high proportion of their income on water. For example, in Port-au-Prince (Haiti) the poorest households sometimes spend 20 percent of their income on water; in Onitsha (Nigeria) the poor pay an estimated 18 percent of their income on water during the dry season versus 2 to 3 percent for the upper-income households; and in Addis Ababa (Ethiopia) and Ukunda (Kenya) the urban poor spend up to 9 percent of their income on water. In Jakarta (Indonesia), of the 7.9 million inhabitants, only 14 percent of households received water through direct connections to the municipal system. Another 32 percent bought water from street vendors who charged about US$1.50 to US$5.20/m^3, depending on the distance from the public tap. In some cases, households purchasing from vendors pay as much as 50 to 60 times more per unit of water than households connected to the municipal system. Some examples of this phenomenon are listed in the following table.

Ratio Between Prices Charged by Vendor and Public Utilities

Country	City	Ratio
Bangladesh	Dacca	12–25
Colombia	Cali	10
Côte d'Ivoire	Abidjan	5
Ecuador	Guayaquil	20
Haiti	Port-au-Prince	17–100
Honduras	Tegucigalpa	16–34
Indonesia	DKI Jakarta	4–60
	Surabaya	20–60
Kenya	Nairobi	7–11
Mauritania	Nouakchott	100
Nigeria	Lagos	4–10
	Onitsha	6–38
Pakistan	Karachi	28–83
Peru	Lima	17
Togo	Lome	7–10
Turkey	Istanbul	10
Uganda	Kampala	4–9

such as the distributions network. In contrast, delivery of services concerns the actual construction, operation, and maintenance of the systems, including monitoring, fee collection, and operation of the treatment facilities, pumping units, and so on. In the past, most of these activities have been conducted by government agencies or public utilities. However, with high levels of unaccounted-for water in many developing country cities such as Manila, there is a growing concern about the inadequacy of water supply systems. Therefore, alternative approaches are being suggested. The poor in many developing countries pay 5 to 100 times more for their water supply than the high-income people with connections to the city water systems (Box 4). Furthermore, even after the International Drinking Water Supply and Sanitation Decade, nearly 1 billion people in the developing world are without access to potable water, and 1.7 billion must contend with inadequate sanitation facilities. The growth in population, combined with funding limitations, poor cost recovery, inadequate operation and maintenance, and lack of trained staff, prevented these efforts from reaching the goals set by the international community at the beginning of the decade.

The public sector will unlikely be able to shoulder the burden of filling the remaining gaps alone—nor is it at all clear that it should try. However, the current attractiveness of demand management tends to be in large part because it has received less attention in recent decades. There are some indications that, especially in the absence of reallocation between uses, returns to demand management will decline, sometimes sharply. As water economies continue to mature, it will be increasingly necessary to formulate ever more sophisticated blending of both supply-side and demand-side options, much as many U.S. electric power utilities already do. This action will place ever greater demands on the mix of institutions governing water use, and in many places will bring to an end the era of "cheap" water.

Note

1. A small portion of this chapter appeared in Nickum and Easter (1990). We are particularly grateful for the wise counsel of Maynard Hufschmidt, Regina Gregory, Helen Takeuchi, and Jennifer Turner in helping us shape the present text, and to Linda Shimabukuro, Loraine Ikeda, and Joyce Kim for making it presentable for publication.

References

Coward, E. Walter, Jr. 1986. "Direct and Indirect Alternatives for Irrigation Investment and the Creation of Property," in K. William Easter, ed., *Irrigation*

Investment, Technology and Management Strategies for Development. Pp. 225–244. Boulder: Westview.

Dani, Anis A. 1986. "Annexation, Alienation, and Underdevelopment of the Watershed Community in the Hindu Kush–Himalayan Region," in K.W. Easter, J.A. Dixon, and M.M. Hufschmidt, eds., *Watershed Resources Management: An Integrated Framework with Studies from Asia and the Pacific.* Pp. 145–158. Boulder: Westview.

Easter, K. William, ed. 1986. *Irrigation Investment, Technology and Management Strategies for Development.* Boulder: Westview.

Easter, K. William, John A. Dixon, and Maynard M. Hufschmidt, eds. 1986. *Watershed Resources Management: An Integrated Framework with Studies from Asia and the Pacific.* Boulder: Westview.

Easter, K. William, and K. Palanisami. 1976. "Tank Irrigation in India: An Example of Common Property Resource Management," in *Proceedings of the Conference on Common Property Resource Management.* Pp. 215–229. Washington, D.C.: National Academy Press.

Grossman, Gregory. 1974. *Economic Systems.* 2nd ed. Englewood Cliffs, N.J.: Prentice-Hall.

Hunt, Robert C. 1985. Appropriate Social Organization? Water Users Association in Bureaucratic Canal Irrigation. Brandeis University, Department of Anthropology, Waltham, Massachusetts.

Korten, Frances F., and Robert Y. Siy, Jr., eds. 1988. *Transforming a Bureaucracy: The Experience of the National Irrigation Administration.* West Hartford, Conn.: Kumarian.

Kumazaki, Minoru. 1982. "Sharing Financial Responsibility with Water Users for Improvement in Forested Watersheds," in *The Current State of Japanese Forestry (II), Its Problems and Future.* Pp. 106–114. Tokyo, Japan: The Japanese Forestry Economic Society.

Letson, David. 1992. "Point/Nonpoint Source Pollution Reduction Trading: An Interpretive Survey." *Natural Resources Journal* 32(2): 219–232 (Spring).

Lovei, Laslo, and Dale Whittington. 1991. Rent Seeking in Water Supply. World Bank Discussion Paper, Report INU85, September.

Nickum, James E., guest ed. 1991. *Water Use Conflicts in Asian-Pacific Metropolises.* Special Issue of *Regional Development Dialogue* 12(4) (Winter).

Nickum, James E. 1993. *Koozooteki na joohoo mondai to kankyoo no kanri* (Structural information problems and environmental management). *Kankyoo Kenkyuu* (Environmental Research Quarterly), No. 90:105–117 (in Japanese).

Nickum, James E., and K. William Easter. 1990. "Institutional Arrangements for Managing Water Conflicts in Lake Basins." *Natural Resources Forum* 14:210–221 (August).

Ostrom, Elinor. 1990. *Governing the Commons.* Cambridge: Cambridge University Press.

Randall, Alan. 1981. "Property Entitlements and Pricing Policies for a Maturing Water Economy." *Australian Journal of Agricultural Economics* 25(3): 195–220.

Revelle, Charles S. 1988. *WSTS Newsletter* (Water Science and Technology Board, National Research Council) 5(3): 1–4 (May).

Runge, Carlisle Ford. 1984. "Institutions and the Free Rider: The Assurance Problem in Collective Action." *The Journal of Politics* 46:154–181.

Saliba, Bonnie Colby. 1987. "Do Water Markets 'Work'? Market Transfers and Trade-Offs in the Southwestern States." *Water Resources Research* 23(7): 1113–1122 (July).

Swaminathan, M. S. 1986. "Foreword," in *Proceedings of the Conference on Common Property Resource Management*. Pp. v–vi. Washington, D.C.: National Academy Press.

United States General Accounting Office. 1989. Reports Issued in May 1989. Washington, D.C.: General Accounting Office. (The report referred to is GAO/RCED–89–101, 25 April, "Water Pollution: Improved Monitoring and Enforcement Needed for Toxic Pollutants Entering Sewers.")

Vermillion, Douglas L. 1991. The Turnover and Self-Management of Irrigation Institutions in Developing Countries. International Irrigation Management Institute, Colombo.

Wolf, Charles, Jr. 1979. "A Theory of Non-Market Failure: Framework for Implementation Analysis." *Journal of Law and Economics* 22:107–139.

World Bank. 1993. Water Resources Management. A World Bank Policy Paper. Washington, D.C.

World Development Report 1992. 1992. New York: Oxford University Press.

Index

About the Book

Although Asia is the least urbanized continent, it contains half of the world's megacities and many of the world's fastest-growing economies. Urban growth is already stressing local water supplies and causing intense conflict among water users—between haves and have-nots in urban areas as well as between farmers and fishers outside the cities. In addition, concern is growing over the depletion and degradation of water sources and over the impact of water policies and patterns of water use on the natural environment.

From the perspective of the maturing metropolitan water economy, the contributors to this volume consider the problems of urban water management in the region. They focus on the institutional and policy dimensions of conflict and seek to provide a range of viable options for reducing the growing frictions among water users. Eight specific case studies of urban areas in Asia and the Pacific span a wide range of economic levels of development, physical settings, and hydrological conditions. The book will be of interest to scholars and policy-makers concerned with issues of water and environmental policy, urban management, and resource conflict in general.